AF738456

ESTUARY

Edited by **William Froneman**

Estuary
http://dx.doi.org/10.5772/66262
Edited by William Froneman

Contributors
William Froneman, Francisco Campuzano, João Sobrinho, Hilda De Pablo, David Brito, Rodrigo Fernandes, Manuela Juliano, Ramiro Neves, Maria João Rocha, Catarina Rocha Cruzeiro, Eduardo Rocha, Aramis Olivos-Ortiz, Carlos A. Zenteno Palma, Sonia I. Quijano Scheggia, María C. Álvarez García, Gloria A. Jiménez Ramón, Ernesto Brugnoli, José Verocai, Pablo Muniz, Felipe Garcia-Rodriguez

Published by InTech
Janeza Trdine 9, 51000 Rijeka, Croatia

© The Editor(s) and the Author(s) 2018
The moral rights of the editor(s) and the author(s) have been asserted.
All rights to the book as a whole are reserved by InTech. The book as a whole (compilation) cannot be reproduced, distributed or used for commercial or non-commercial purposes without InTech's written permission. Enquiries concerning the use of the book should be directed to InTech's rights and permissions department (permissions@intechopen.com).
Violations are liable to prosecution under the governing Copyright Law.

(cc) BY

Individual chapters of this publication are distributed under the terms of the Creative Commons Attribution 3.0 Unported License which permits commercial use, distribution and reproduction of the individual chapters, provided the original author(s) and source publication are appropriately acknowledged. More details and guidelines concerning content reuse and adaptation can be found at http://www.intechopen.com/copyright-policy.html.

Notice
Statements and opinions expressed in the chapters are these of the individual contributors and not necessarily those of the editors or publisher. No responsibility is accepted for the accuracy of information contained in the published chapters. The publisher assumes no responsibility for any damage or injury to persons or property arising out of the use of any materials, instructions, methods or ideas contained in the book.

Publishing Process Manager Dajana Pemac
Technical Editor SPi Global
Cover InTech Design team

First published March, 2018
Printed in Croatia
Legal deposit, Croatia: National and University Library in Zagreb

Additional hard copies can be obtained from orders@intechopen.com

Estuary, Edited by William Froneman
p. cm.
Print ISBN 978-953-51-3879-2
Online ISBN 978-953-51-3880-8

We are IntechOpen, the first native scientific publisher of Open Access books

3,350+
Open access books available

108,000+
International authors and editors

114M+
Downloads

151
Countries delivered to

Our authors are among the
Top 1%
most cited scientists

12.2%
Contributors from top 500 universities

WEB OF SCIENCE™

Selection of our books indexed in the Book Citation Index
in Web of Science™ Core Collection (BKCI)

Interested in publishing with us?
Contact book.department@intechopen.com

Numbers displayed above are based on latest data collected.
For more information visit www.intechopen.com

For EU product safety concerns:
IN TECH d.o.o., Prolaz Marije Krucifikse Kozulić 3, 51000 Rijeka, Croatia,
info@intechopen.com or visit our website at intechopen.com.

Meet the editor

William Froneman, PhD, is currently the head of the Department of Zoology and Entomology and the director of the Southern Ocean Group at Rhodes University, South Africa. His research focuses on understanding the spatial and temporal dynamics of aquatic food webs in both shallow (ephemeral ponds and estuaries) and deep water systems. Since obtaining his PhD degree in 1996, he has published 172 peer-reviewed science journal articles and 10 book chapters and has successfully supervised 38 MSc theses and PhD dissertations. In recognition of his research achievements, he received several awards including the Junior and Senior Distinguished Research Awards from Rhodes University and the Meiring Naude Gold Medal from the Royal Society of South Africa.

Contents

Preface

Estuaries form a transition zone between river and maritime environments and, as such, are subjected to both marine and riverine influences. Estuaries are regarded among the most ecologically threatened ecosystems worldwide largely due to anthropogenic impacts including freshwater abstraction, habitat loss, and the overexploitation of resources. More recently, these systems have been identified as important repositories for various contaminants including heavy metals and pesticides. The development and successful implementation of management and conservation strategies to conserve these ecologically threatened ecosystems are critically dependent on understanding the links among hydrological, physicochemical, and biological variables, and these interactions are likely to be altered with global change. The book provides a comprehensive overview of selected topics including modeling of water exchange between estuaries and the ocean, sediment geochemistry and mangrove health, climate variability and hydrology, and pesticides in estuaries and ecosystem functioning for various estuaries including permanently open, mangrove, and intermittently open/closed systems in both the northern and the southern hemispheres.

Chapter 1. *Coupling watersheds, estuaries, and regional oceanography through numerical modeling in the Western Iberia: thermohaline flux variability at the ocean-estuary interface.*

The characterization of water and property exchanges at the estuary-ocean interface is key information to understand the estuarine plume influence on coastal circulation and in the generation of haline fronts. In this chapter, water and property exchanges at the estuary-ocean interface for the eighth largest Portuguese estuaries were modeled using the MOHID water numerical model for the period of 2010–2016. The numerical analysis of the estuarine fluxes allowed for the better characterization of the various systems with contrasting catchment characteristics. Where available, the modeling results were compared to field data collected near the estuary mouth.

Chapter 2. *Weather, hydrological, and oceanographic conditions on the northern coast of the Rio de la Plata estuary during the 2009–2010 ENSO.*

Climatic, hydrological, and oceanographic conditions were determined during the 2009–2010 El Nino-Southern Oscillation (ENSO) on the north coast of the Rio de la Plata (RdlP) estuary. Monthly flow of RdlP demonstrated significant differences between the ENSO phases. Minimum wind stress was recorded during the 2009–2010 ENSO, during which time the maximum flow of RdlP was observed. The high flow contributed to variations in salinity, water column stratification, nutrient availability, and light penetration, which had a marked impact on the biology of the system. The spatial location of the turbidity front was associated with the flow of RdlP and, to a lesser extent, wind stress.

Chapter 3. *Pesticides in worldwide aquatic systems—Part I*

The occurrence of pesticides in aquatic environments is a worldwide phenomenon. In this chapter, data (the number of quantified pesticides, pesticide category, minimum, maximum, and average amounts) collected from 95 aquatic systems worldwide over the past 17 years were analyzed. Data were evaluated by continent, although the focus of the study was Europe where the largest number of studies has been conducted. Insecticides were the most common category of pesticides often used in excess on the Asian continent. Moreover, priority pesticides settled for elimination (Stockholm Convention list) were still present in almost all the continents, demonstrating that years after the last convention, these compounds continue to be employed. This contributed to aquatic systems containing both legal and illegal pesticides, which have adverse effects on different trophic levels, including humans. The study suggests that global effort is required to regulate the use of pesticide and that continuous environmental monitoring should be enforced to understand the potential toxicological risks.

Chapter 4. *Pesticides in worldwide aquatic systems—Part II*

The contamination of aquatic systems by pesticides is a worldwide problem that can have an adverse environmental impact. Chapter 4 presents the main findings of a desktop study that assessed possible environmental risks through theoretical calculations, using worldwide data published over the past 17 years taking into account different trophic levels (algae, invertebrates, and fish) and the maximum average environmental concentrations (in water) observed in each continent. Furthermore, hazard quotients—using the estimated average daily intake, theoretical maximum daily intake, and the maximum residue limits—were calculated to estimate the potential risks to humans through the consumption of molluscs, crustaceans, and fish. Data indicate that algae and invertebrates are the most sensitive groups to pesticide contamination. Elevated concentrations of pesticides were recorded in fish suggesting that the consumption of fish may pose a health threat.

Chapter 5. *The ecology and food web dynamics of the southern African intermittently open/closed estuaries*

Chapter 5 provides an overview of the ecology and food web dynamics of the southern African intermittently open/closed estuaries (IOCEs) and highlights the importance of freshwater inflow, mouth phase (open versus closed), and the establishment of a link to the marine environment in determining the trophic interactions and community composition within these systems. The potential impact of global climatic change on ecosystem functioning of these systems is discussed.

Chapter 6. *The relationship of sediment and interstitial water properties with mangrove health in a subtropical coastal lagoon of Mexico.*

The influence of the physicochemical properties of interstitial water, sediment geochemistry, and sediment grain size on the health of mangroves was investigated at eight stations in the Cuyutlán lagoon, Mexico. The results of the study indicate that mangrove health was strongly determined by tidal influence, which contributed to an increase in dissolved oxygen and organic carbon concentrations and a decrease in salinities.

William Froneman, PhD
Head of Department of Zoology and Entomology
Director of the Southern Ocean Group
Rhodes University, South Africa

Chapter 1

Coupling Watersheds, Estuaries and Regional Oceanography through Numerical Modelling in the Western Iberia: Thermohaline Flux Variability at the Ocean-Estuary Interface

Francisco J. Campuzano, Manuela Juliano,
João Sobrinho, Hilda de Pablo, David Brito,
Rodrigo Fernandes and Ramiro Neves

Additional information is available at the end of the chapter

http://dx.doi.org/10.5772/intechopen.72162

Abstract

The characterisation of the water and properties exchanges at the estuary-ocean interface is a key information to understand the estuarine plume influence on coastal circulation and in the generation of haline fronts. In this work, the largest eight Portuguese estuaries were modelled using the MOHID Water numerical model for the period 2010–2016. Water fluxes and associated properties were computed numerically at each of the estuary mouths. These results served to estimate the tidal prisms, tidal flows and to describe the annual evolution of water temperature and salinity. Those fluxes could serve to improve the land boundary conditions for regional ocean models. Moreover, the numerical analysis of the estuarine fluxes allow for the better characterisation of the studied systems, as two neighbouring estuaries could present very different fluxes and water properties. Where available, modelling results were compared with stations near the estuary mouth.

Keywords: Portugal, numerical modelling, MOHID, PCOMS, WIBP, salinity, estuarine fluxes, tidal prism, river

1. Introduction

In Ref. [1], an estuary is defined as 'a semi-enclosed coastal body of water, which has a free connection with the open sea, and within which sea water is measurably diluted with freshwater derived from land drainage'. On its way to the open ocean, fresh water carried by rivers

INTECH
open science | open minds

© 2018 The Author(s). Licensee InTech. This chapter is distributed under the terms of the Creative Commons Attribution License (http://creativecommons.org/licenses/by/3.0), which permits unrestricted use, distribution, and reproduction in any medium, provided the original work is properly cited. (cc) BY is mixed with the saltier ocean water aided mainly by the action of tides and atmospheric forces. The salinity of the estuarine fluxes that are incorporated in the near ocean depend on several factors including the estuary tidal prism and magnitude of river run-off. While the size of the tidal prism is controlled by the morphodynamics of each particular area, the fresh water inputs are controlled by natural factors such as rainfall and human fresh water management upstream the river mouth, i.e. water irrigation, storage for electricity and deviation to other watersheds. The natural and man-made variability of the fresh water volume and properties is a constraint to characterise the estuarine waters that flow from estuaries to the open ocean due to the lack of permanent monitoring stations. This characterisation is critical to understand their influence in coastal hydrodynamics and ecological processes.

From the European environmental legislation perspective, estuaries are typically regarded as transitional waters in the EU Water Framework Directive (WFD, Directive 2000/60/EC) and therefore may mostly fall outside the scope of EU Marine Strategy Framework Directive (MSFD, Directive 2008/56/EC). However, estuaries may have estuarine plumes which extend beyond transitional waters limit, 1 nautical mile from baseline, into coastal waters or beyond; in these cases, the estuarine habitat would fall within the scope of MSFD [2]. For this reason, the exclusion of estuaries and other transitional waters from the MSFD has been questioned [3]. In any case, though WFD requires the monitoring of several biological and chemical elements for the establishment of reference conditions in terms of water quality, long-term continuous monitoring with high temporal resolution are rarely deployed in estuaries [4] and, when deployed, some sensors have limited capabilities for longstanding observations [5]. For this reason, numerical models are able to support *in situ* monitoring by completing the datasets spatially and temporally and by contrasting the ambiguous observed values. Also numerical models can assist to describe the transitional waters including the extension and influence of the estuarine plumes in the near ocean.

In this chapter, the tidal prism and estuarine fluxes will be characterised with the aid of numerical models and the existing monitoring stations from the main Portuguese estuaries. This work also aims to describe in more detail the estuarine component of the methodology to couple watersheds, estuaries and regional oceans through numerical modelling for Western Iberia [6].

2. Study area

In this study, fluxes from the eight largest Portuguese estuaries were analysed. These were from North to South: Minho, Lima, Douro, Ria de Aveiro (hereafter referred as Aveiro estuary), Mondego, Tagus, Sado and Guadiana (**Figure 1**). All these estuaries are named after the main river discharging into the system with the exception of the Aveiro Estuary, where the principal fresh water source is the Vouga River. Their associated river catchments are distributed in areas with different climate characteristics; for example, the annual rainfall gradient between the two most distant rivers, Minho and Guadiana, which is around 1700 mm [7]. From an administrative point of view, Douro, Guadiana, Lima, Minho and Tagus are international rivers while the drainage areas of the Vouga, Mondego and Sado rivers fall entirely within Portuguese territory.

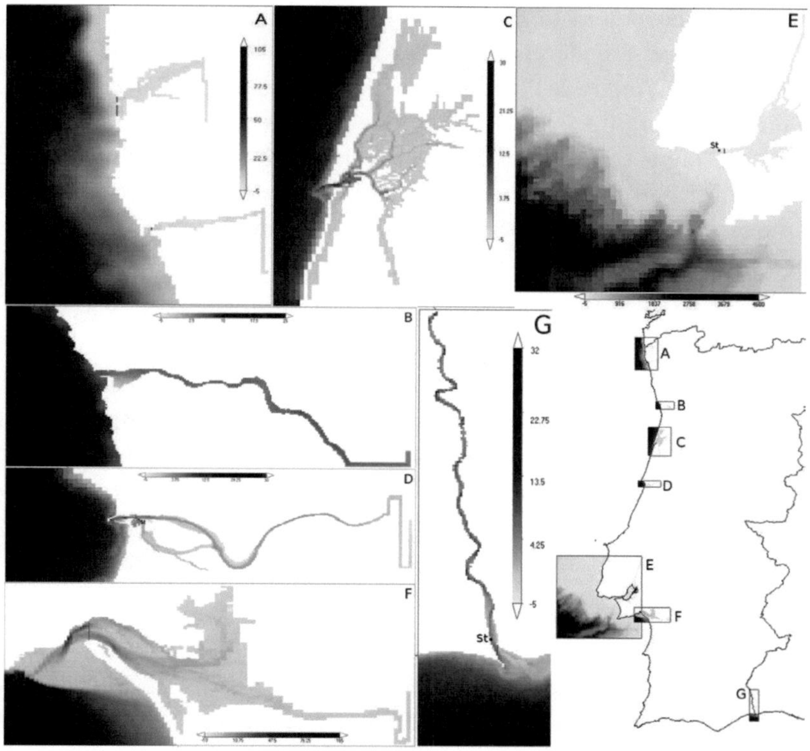

Figure 1. Bathymetry and model domain of the estuaries analysed in this work and its location in the Portuguese context (bottom right): (A) Minho (top) and lima (bottom), (B) Douro, (C) Aveiro, (D) Mondego, (E) Tagus, (F) Sado and (G) Guadiana. In each estuary, a red line indicates the location of the cross-section where the estuarine fluxes were calculated. Monitoring stations in the Guadiana, Mondego and Tagus estuaries are indicated with 'St'.

3. Material and methods

3.1. Monitoring networks

In order to validate the estuarine numerical model applications, automatic data collected along the estuarine continuum from hydrometric stations and estuarine buoys were gathered and analysed during this investigation. Hydrometric data provided reliable boundary conditions for the river, while the *in situ* estuarine data collected by the monitoring stations were used to validate the modelling results. Although influenced by the global decline context [8], the hydrometric network is a more stable and reliable source of data when compared to the long-term autonomous monitoring of estuarine systems which is generally scarce. In fact, none of the estuarine stations described below are currently in operation. Moreover, to guarantee high performance, the sensors equipped in the estuarine stations require periodic calibration [4]. For instance, the performance of temperature sensors are more reliable than salinity sensors [5].

The Portuguese river hydrometric observation network, forms part of the Portuguese Environmental Agency (APA, by its acronym in Portuguese), and is responsible for monitoring the fresh water flow in all the territory and data is made freely available in the SNIRH website (http://snirh.pt). Reliable automatic stations for the study period were available in the Douro, Guadiana, Mondego and Tagus rivers and are Albufeira de Crestuma, 20 km upstream from Douro estuarine mouth, Açude de Pedrogão, 130 km upstream from Guadiana estuary mouth, Açude de Coimbra, 45 km upstream from Mondego estuary mouth, and Almourol, 130 km upstream from the Tagus estuary mouth. The location of these stations will exclude the contributions to the river flow due to rainfall and other minor tributaries between the station and the head of the estuary, which can be relevant in heavy rainy areas.

In Portugal, several estuaries were monitored continuously during the simulated period, 2010–2016. The SIMPATICO network comprising three *in situ* monitoring stations in the Guadiana (37.188°N, 7.411°W), Mondego (40.144°N, 8.856°W) and Tagus (39.058°N, 8.774°W) estuaries collected continuous water-quality and current data [4]. For the purpose of this study, only the Guadiana and Mondego observations from the SIMPATICO network are adequate, since the monitoring station in the Tagus estuary is located 40 km upstream from the Tagus mouth and thus unable to represent the estuarine mouth conditions. Instead, a monitoring buoy, hereafter referred as Algés buoy (38.694°N, 9.237°W), deployed by the water utility SimTejo was used to represent the Tagus water conditions in the estuary-ocean interface (**Figure 1**).

3.2. Numerical models

Each studied estuary was simulated for the period 2010–2016 using the MOHID Water model. The MOHID Water model is an open-source numerical model included in MOHID Water Modelling System (http://www.mohid.com; [9]) developed continuously since 1985 mainly at the Instituto Superior Técnico (IST) from the Universidade de Lisboa. The model adopted an object oriented philosophy model for surface water bodies that integrates different scales and processes using finite volumes. The core of the model is a fully 3D hydrodynamic model which is coupled to different modules comprising among others water quality, atmosphere processes, discharges, oil dispersion, mixing zone model for point source discharges. The MOHID Water model has been applied to many coastal and estuarine areas worldwide and has shown its ability to simulate successfully very different spatial scales from large coastal areas to coastal structures.

For this study, most of the estuarine modelling applications were simulated using modelling grids, domains and bathymetries from previous studies (**Table 1**). The bathymetries of the Minho, Lima and Sado estuarine model applications were updated on their ocean side using improved global bathymetric databases such as the EMODnet-Hydrography portal (http://www.emodnet-hydrography.eu). Only the Guadiana estuary model domain was fully developed for this study, using a 100 m resolution bathymetry model for the Guadiana estuary provided by the Instituto Hidrográfico (http://www.hidrografico.pt).

Boundary conditions were updated for most of the applications; on the ocean side, the estuarine models were nested to the Portuguese Coast Operational Modelling System (hereafter referred as PCOMS; [10]), following the downscaling methodology described in Ref. [11]. On the landward side, Douro, Guadiana, Mondego and Tagus estuaries were forced by river flow data obtained from the closest stable monitoring station from the Portuguese river hydrometric observation network (http://snirh.pt; **Table 1**). Due to the limited data available for the rivers discharging in the Aveiro, Lima, Minho and Sado estuaries, flow data were obtained from a

Estuary	Mouth Location		Horizontal resolution	Dimension (cells number)		Meteorology model	Fresh water sources	References
	Lat.	Long.		Zonal	Meridional			
Aveiro	40.645°N	8.740°W	0.0022*	87	81	MM5 IST-9 km	**Model:** Vouga MOHID Land **Climatology:** Antuã, Boco, Caster and Mira rivers	Saraiva [17] and Sobrinho [18]
Douro	41.146°N	8.671°W	0.0010	244	115	WRF MG-4 km	**SNIRH:** Albufeira de Crestuma station	Kenov I. (Pers. Comm.)
Guadiana	37.178°N	7.405°W	0.0018	70	221	MM5 IST-9 km	**SNIRH:** Pedrogão station	developed for this study
Lima	41.686°N	8.835°W	0.0040	70	105	WRF MG-4 km	**Model:** Lima MOHID Land	updated from Saraiva et al. [19]
Minho	41.858°N	8.875°W					**Model:** Minho MOHID Land	
Mondego	40.145°N	8.866°W	0.0035	206	97	MM5 IST-9 km	**SNIRH:** Açude de Coimbra station	Kenov et al. [20]
Sado	38.503°N	8.905°W	0.0060	94	90	MM5 IST-9 km	**Model:** Sado MOHID Land	updated from Martins et al. [21]
Tagus	38.685°N	9.222°W	0.0027*	145	120	WRF IST-3 km	**SNIRH:** Almourol station **Climatology:** Sorraia and Trancão rivers **Constant:** UWWTPs of Alcântara, Beirolas, Chelas, S. João da Talha and Guia	Fernandes [22] and Campuzano et al. [11]

Cross-section estuarine fluxes are calculated along the defined longitude while latitude indicates the centre of the mouth section. Observations were obtained from the SNIRH web portal (http://www.snirh.pt) and model results were obtained from the West Iberia domain (WI MOHID Land model application) described in Ref [12].
*The zonal resolution at the mouth in irregularly sized grids.

Table 1. Estuary mouth coordinates, horizontal resolution, model domain dimension, sources of meteorological forcing and fresh water and previous model description reference.

watershed model for West Iberia with 2 km horizontal resolution (WI MOHID Land) described in Ref. [12]. Monthly climatology river flow or constant values were only used for other minor sources of fresh water, the Pranto river for the Mondego estuary, four for the Aveiro estuary and two rivers and 14 urban waste water treatment plants (UWWTPs) for the Tagus estuary. As monitoring stations providing reliable and regular water properties were absent for many of the rivers, water temperature in all the estuarine applications was obtained from the WI MOHID Land model application and river salinity was considered as pure fresh water.

All the estuarine models were implemented in 2D depth integrated domains except for the Tagus application, hereafter referred as TagusMouth, which is a fully 3D baroclinic hydrodynamic and ecological application. The TagusMouth vertical discretisation consists of a mixed vertical geometry which is composed of a sigma domain with 7 layers from the surface until 8.68 m depth, with variable thickness decreasing up to 1 m at the surface, on top of a Cartesian domain of 43 layers with thickness increasing towards the bottom.

Each estuarine application was forced with the higher horizontal resolution meteorological model available for its domain. Meteorological model applications provided 3D fields that included relevant model forcing variables (i.e. precipitation, solar radiation, wind modulus and direction, relative humidity, air temperature, etc.) and whose surface layer is interpolated for each modelling domain by triangulation. The Minho, Lima and Douro estuaries were forced using WRF model results [13] with 4 km horizontal resolution generated by Meteogalicia (MG, http://www.meteogalicia.es). The TagusMouth model application is forced using 3 km horizontal resolution WRF model results provided by the IST meteorological group (IST, http://meteo.tecnico.ulisboa.pt). The Aveiro, Guadiana, Mondego and Sado modelling domains used, as atmospheric boundary conditions, MM5 modelling results [14] with a horizontal resolution of 9 km provided also by the IST meteorological group [15].

This set of numerical models was integrated and synchronised through the ART software (Automatic Running Tool; **Figure 2**), a software for the automation of model simulations developed at IST that is currently used by many operational applications [6, 12, 16]. The ART tool is a standalone application, independent, compiled, and able to run in any Windows operative system. This tool can be seen as the 'heart' of an operational framework, controlling the execution of other auxiliary standalone applications or even scripts (e.g. conversion of file formats, interpolation and specific downloading procedures), adapting automatically the configuration files and launching those applications. The ART tool pre-processes the boundary conditions from different sources needed to run the model, and executes the MOHID Water and Land model applications using the configured files. After each estuarine model has concluded, it executes the River Out software, a software designed specifically to extract and calculate the water fluxes and concentrations for all the cells comprised at a fixed longitude and between two defined latitudes and to store them in an organised manner.

Table 1 summarises the main characteristics of the estuarine model applications including the modelling domains, their location of their mouth where the fluxes were calculated, the source of fresh water forcing and the reference of the previous study. Most of the numerical models here described are currently running operationally in the MARETEC research centre (http://forecast.maretec.org).

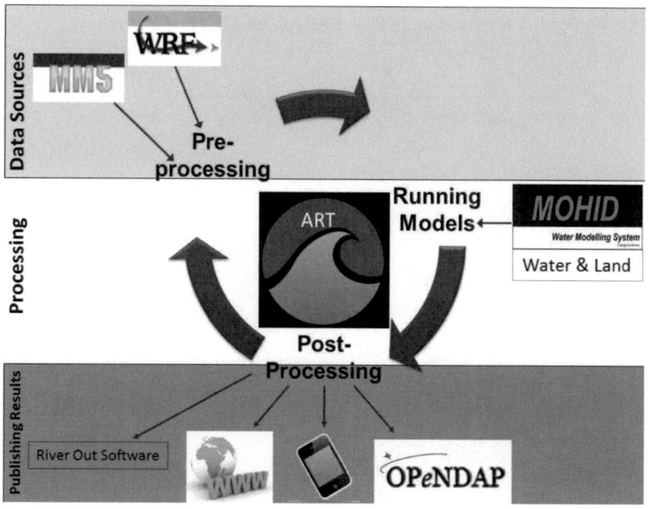

Figure 2. General scheme of the Automatic Running Tool (ART) depicting the pre-processing, modelling and post-processing cycle of operations with the elements used for the MOHID Land and MOHID Water applications used in this work. The River Out software is executed during the post-processing operations of the estuarine model application.

4. Results and discussion

In Portuguese estuaries, the ocean and fresh water mix are controlled by the tidal regime, estuary geomorphology and river discharge. A useful parameter to characterise the estuaries by the tidal regime and the geomorphology is the tidal prism. The tidal prism, as defined by [23], is 'the amount of water that flows into and out of an estuary or bay with the flood and ebb of the tide, excluding any contribution from freshwater inflows'.

The defined cross-sections serve to estimate numerically and accurately the amount of water flowing through the estuarine mouth and thus to estimate tidal prisms. Tidal prism volumes were calculated by integrating the hourly ebb and flood flows for each tidal cycle. For each estuary, mean tidal prism values were obtained considering the entire dataset and differentiating between ebb and flood conditions obtaining volumes with a similar order of magnitude than the ones found in the literature (**Table 2**).

The volume of water discharged by the river affects the duration and intensity of the flood and ebb tides. Ebb tidal prisms are generally larger as they include the river flow in addition to the flushing tide. Furthermore, high river discharges constraint the beginning of the estuary flooding. The difference between flood and ebb tidal prisms is larger in estuaries with low tidal prism and high river discharge such as that in the Douro estuary. When the Douro river flow is removed from the fluxes calculation, the obtained tidal prisms present similar volumes for both flood and ebb tides, estimated at around 1.21×10^7 m^3. As can be seen in **Table 2**, the Douro estuary can be defined as an extreme case for Portuguese estuaries, as in the remaining systems, the rivers do not appear to influence the tidal prism and differences between ebb and flood are similar. For instance, the tidal prism of the Ria de Aveiro without the Vouga river discharge was 6.81×10^7 m^3 which is close to the mean tidal prism volume.

Estuary	Mean TP (×10⁷ m³)	Flood Mean TP (×10⁷ m³)	Ebb Mean TP (×10⁷ m³)	Flow std. deviation (σ) (m³ s⁻¹)	TP Reference volume (×10⁷ m³)
Aveiro	6.83 ± 1.82	6.61 ± 1.86	7.04 ± 1.76	3518.92	6.00 [24] 7.00 [25] 11.90 [26]
Douro	2.20 ± 12.23	0.83 ± 0.49	3.56 ± 23.97	875.99	0.55 [27] 2.09 [26]
Guadiana	2.96 ± 1.74	2.73 ± 0.74	3.19 ± 2.73	1467.16	0.60 [26]
Lima	1.24 ± 0.57	1.08 ± 0.52	1.40 ± 0.61	749.74	0.90 [26]
Minho	3.10 ± 1.53	2.79 ± 1.39	3.41 ± 1.66	1612.07	0.55 [26]
Mondego	1.53 ± 0.86	1.41 ± 0.52	1.64 ± 1.10	802.58	0.99 [28]
Sado	29.93 ± 7.98	29.83 ± 7.95	30.02 ± 8.00	15,064.64	40.00 [29]
Tagus	62.47 ± 15.96	62.01 ± 16.50	62.93 ± 15.42	31,753.82	75.00 [30]

The table also included the estuarine flow standard deviation considered as an indicator of the flow at the ocean-estuary interface.

Table 2. Tidal prism (TP) for the entire study period and distinguishing between flood and ebb periods for each of the estuarine systems and bibliographic reference volumes.

As the estuarine fluxes changes during ebb and flood conditions, the net flux is similar to the river flow and always towards the ocean. However, the discharge momentum is far more intense than the river flow and evolves over time following the spring and neap tidal cycles. In order to provide a statistical indicator of the exchanged flow with the ocean, the standard deviation (σ) can be employed as an indicator of the mean volume exchange at the estuarine mouth and ±2σ would result in the average maximum volume that circulates through the estuarine-ocean connection (**Table 2**). This value allows one to determine the magnitude of the exchanged flow and the influence in the neighbouring ocean circulation. The Tagus estuary exchange flow is larger than all the other estuaries combined (around 24,000 $m^3 s^{-1}$), followed by the Sado and Aveiro with half and a tenth of the Tagus flow, respectively.[1] On average, the total flow from the eight estuaries to the neighbouring ocean is around 55,000 $m^3 s^{-1}$.

Depending on the tidal prism and the volume of river discharge, the properties of the water exchanged with the coastal area fluctuates. **Figures 3** and **4** indicate the evolution of the mean modelled temperature and salinity in each of the estuaries during the study period. Temperature and salinity values shown for the Tagus estuary in these figures correspond to its surface layer and therefore can be directly compared with the observations. In the rest of model applications, since they are bidimensional, temperature and salinity represent the vertically integrated values for the entire water column.

Surface water temperatures in the Atlantic Ocean around Portugal show a clear seasonal pattern with maximum temperatures during the summer months (JJA) and lower temperatures in winter (DJF) (**Figure 3** and **Table 4**). In tidally dominated estuaries, i.e. large tidal prism, temperature variations are reduced as ocean surface temperatures fluctuate less. For example, the water fluxes temperature for the Sado and Tagus estuaries range between 10 and 25°C, while in

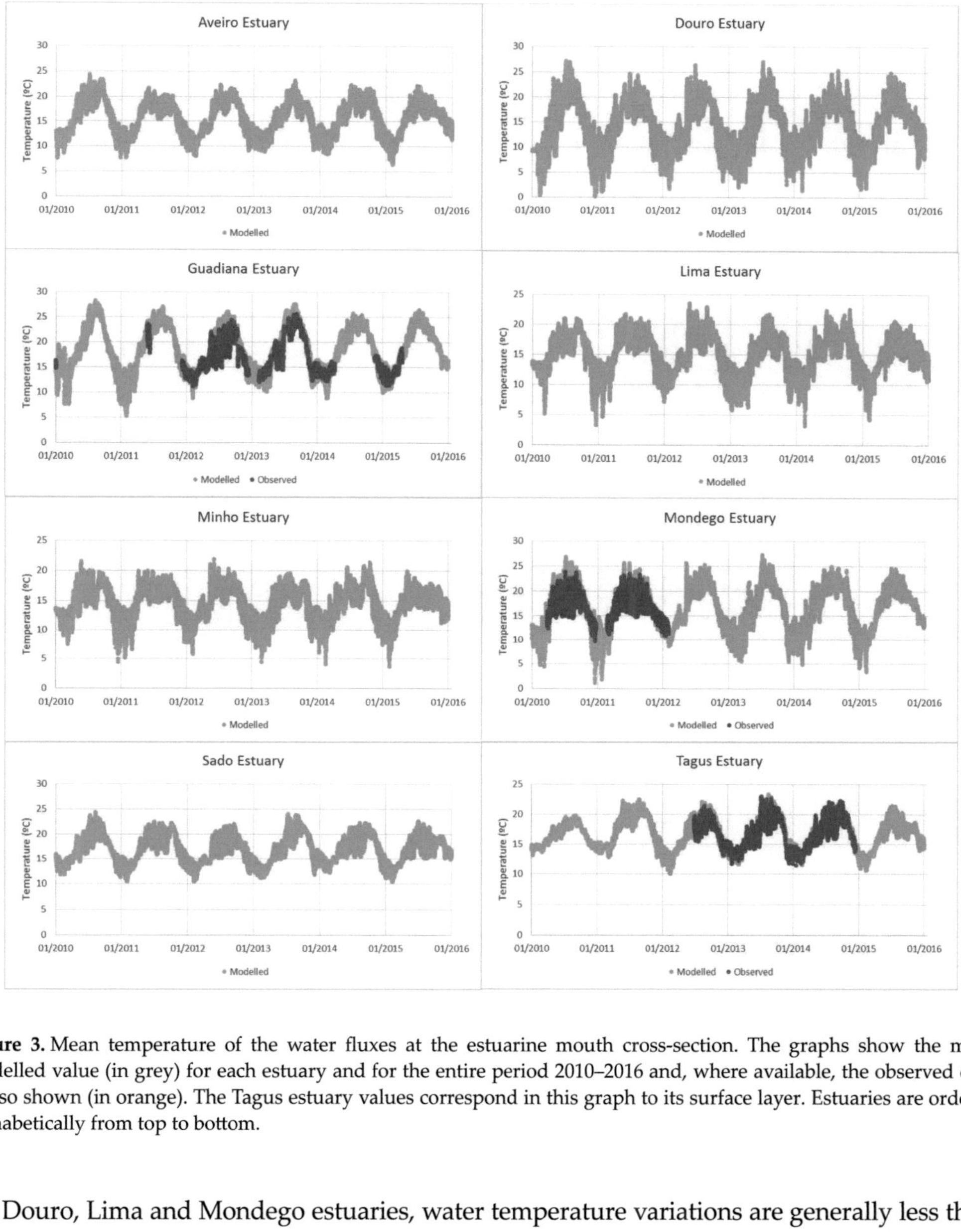

Figure 3. Mean temperature of the water fluxes at the estuarine mouth cross-section. The graphs show the mean modelled value (in grey) for each estuary and for the entire period 2010–2016 and, where available, the observed data is also shown (in orange). The Tagus estuary values correspond in this graph to its surface layer. Estuaries are ordered alphabetically from top to bottom.

the Douro, Lima and Mondego estuaries, water temperature variations are generally less than 5°C. The river effect in the fluxes temperature is more recognisable during the winter period (DJF), as during this season, inland temperatures are lower and the river flow is higher.

The impact of river flow on the estuarine fluxes is most pronounced for salinity and is consistent with the different types of estuary (**Figure 4**). Estuaries with large tidal prisms, such as Sado and Tagus, present salinity concentrations greater than 0 even during high river flows such as the Tagus flooding event of April 2013. On the other hand, the Douro estuary exchanges fresh water with the near ocean on multiple occasions during every season and as a consequence, its fluxes present a mean salinity around 11 (**Table 5**). Guadiana, Lima and Mondego estuaries

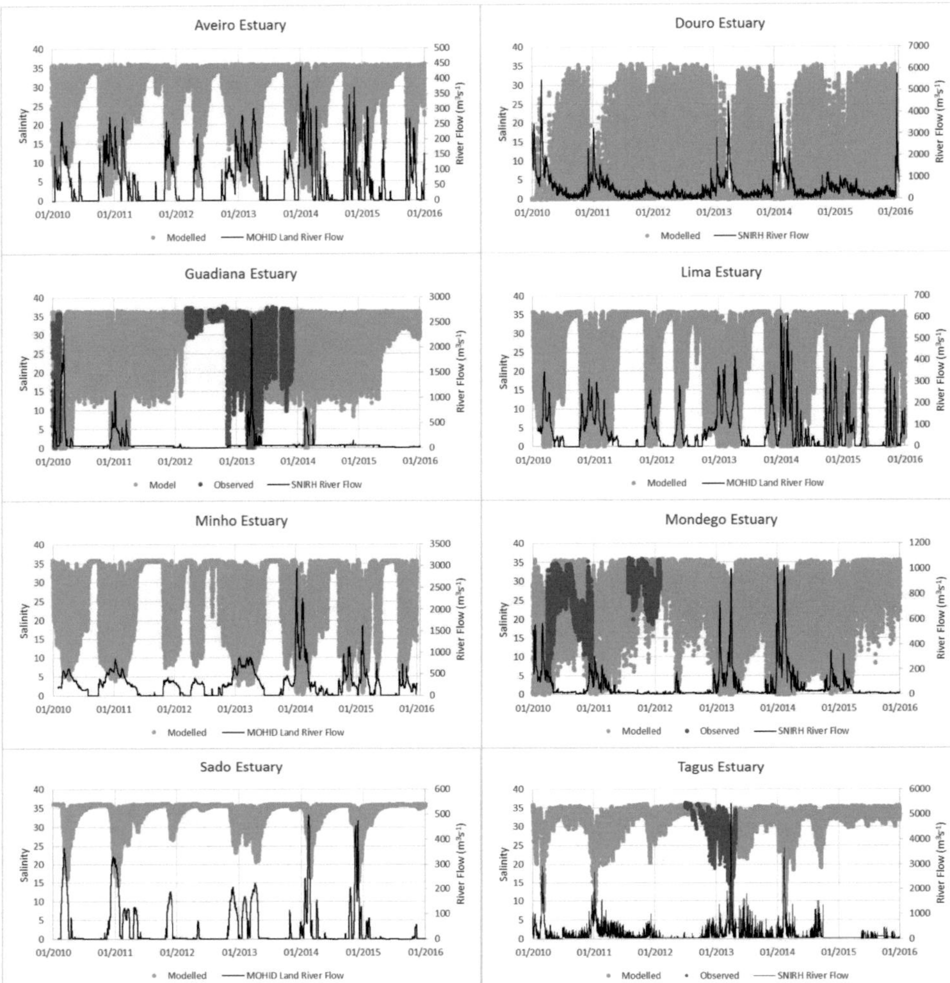

Figure 4. Mean salinity of the water fluxes at the estuarine mouth cross-section. Graphs show the mean modelled value (in grey) for each estuary and for the entire period 2010–2016 and, where available, the observed data is also shown (in orange). The Tagus estuary values correspond in this graph to its surface layer. The river flow imposed in the model is displayed with a continuous black line and a secondary axis value and indicated whether the sources are observations (SNIRH) or watershed model results (MOHID land). Estuaries are ordered alphabetically from top to bottom.

exchange fresh water fluxes periodically every winter period only with the exception of year 2012 which was an extremely dry year. **Figure 4** shows the spatial heterogeneity of rain patterns since high river flows and events are not simultaneous in all the studied estuaries.

As mentioned above, three estuaries, Guadiana, Mondego and Tagus, had monitoring devices near the mouth of the estuary during the study period. The Algés buoy was monitoring near the Tagus estuary mouth for the period 27/06/2012 to 12/12/2014. The SIMPATICO station for the Mondego estuary collected data between 30/10/2007 and 02/02/2012 while in the Guadiana estuary, data was available for the period 19/03/2008 to 21/04/2015. The SIMPATICO dataset collected at the Guadiana estuary until April 2014 [5], is publicly available in machine-readable format at PANGAEA (DOI: 10.1594/PANGAEA.845750).

As discussed earlier, temperature sensor performance is more reliable than salinity sensors during long-term deployment. In that sense, the Guadiana estuary valid temperature data were 99%, while salinity valid data were 67.1% [5] with 20% of invalid data and 12.9% of ambiguous data. Also in the Tagus estuary, the Algés buoy salinity values were identified as valid until 30/04/2013 after which time the sensor calibration was lost. Valid temperature and salinity data for the three monitoring stations are represented in **Figures 3** and **4**. When compared with the modelled data, the coefficient of determination for temperature is higher than that for salinity at each station and the best model performance were for the Guadiana and Tagus estuaries (**Table 3**). The Mondego estuary coefficient of determination is lower in the SIMPATICO station compared with the other two estuaries. A possible reason is its location in the entrance of the southern branch ([4]; **Figure 1**) that could also be influenced by the Pranto River, which was simulated using a river flow climatology and may not be so representative of the main fresh water entrance that take place in the northern branch. It should also be noted that this is a comparison between the observed data at the surface with depth-integrated modelling results.

From the modelling results, a preliminary climatology of water temperature and salinity in the ocean-estuary interface can be obtained, which illustrate their variability along the year and provide a global picture of the thermohaline fluxes from the Portuguese estuaries. **Tables 4** and **5** list the monthly averaged temperature and salinity, respectively, calculated by the model at the ocean-estuary interface for the period 2010–2016 (**Figure 5**). February is the month presenting coldest temperatures for all the estuaries, while the warmest months occur between July (i.e. Douro) and October (i.e. Minho). The estuary location influences the summer maximum temperature as western Iberia coast is subjected to upwelling events during summer that bring cold nutrient-rich waters to the ocean surface. Thus, winter temperatures provide a more accurate reflection of the river discharges contribution to the estuarine water as it coincides with the rainy season. As a result, it can be concluded that the salinity concentrations of the Douro and the Tagus estuaries are, on average, the most and least influenced by river discharge, respectively.

Salinity is a more conservative indicator to study the river effect, as the salinity values in the near ocean vary little and is unaffected by the upwelled waters. Fluxes from Sado and Tagus estuaries could be categorised as saline because monthly averaged salinities were always >30. Generally, the Aveiro estuary exchanges saline waters during summer and autumn months and brackish waters, with salinities between 5 and 30, during winter and spring months while the Lima, Minho and Mondego estuarine mouths are dominated by brackish waters year round. Fresh water, (salinities <5), was recorded at the mouth of the Douro estuary during three first months of the year and, on average, maximum salinity observed during August is

Estuary	Temperature R^2	Salinity R^2
Guadiana	0.90 (N 21,435)	0.74 (N 12,057)
Mondego	0.79 (N 14,741)	0.32 (N 14,741)
Tagus	0.92 (N 20,944)	0.83 (N 7377)

The number of valid values is indicated in brackets.

Table 3. Coefficient of determination (R^2) between the modelling results and the available observed data for the Guadiana, Mondego and Tagus estuaries.

Month	Aveiro	Douro	Guadiana	Lima	Minho	Mondego	Sado	Tagus
January	11.82	9.01	13.44	11.16	11.60	11.57	13.68	13.72
February	11.28	8.42	12.83	10.83	10.92	11.15	12.96	13.07
March	12.76	10.76	13.86	12.36	12.18	12.83	14.12	13.98
April	14.96	13.97	16.60	14.86	14.17	15.69	15.86	15.63
May	16.67	15.84	19.19	16.42	15.37	17.82	17.29	17.04
June	18.07	18.03	21.51	17.58	16.16	19.65	18.56	18.10
July	18.51	19.38	23.81	17.83	16.17	21.07	19.57	18.76
August	18.56	19.16	24.59	17.51	15.88	21.24	19.61	19.20
September	18.89	18.74	23.77	18.06	16.86	20.97	19.96	19.77
October	18.04	16.74	21.29	17.15	17.12	19.15	19.59	19.31
November	15.08	12.27	17.60	13.67	14.50	15.23	17.07	17.17
December	12.96	9.87	14.79	11.86	12.60	12.38	14.75	14.92
Average	15.66	14.38	18.64	14.96	14.48	16.66	16.94	16.81
Range	7.61	10.96	11.76	7.23	6.20	10.09	7.00	6.70

The average value and the range of values are also indicated.

Table 4. Mean monthly modelled temperature for eight analysed estuaries calculated at the ocean-estuary defined cross-sections during the 2010–2016 period.

Month	Aveiro	Douro	Guadiana	Lima	Minho	Mondego	Sado	Tagus
January	26.91	5.12	25.08	15.45	20.12	15.88	32.61	32.20
February	27.41	4.81	24.36	14.81	18.92	17.24	32.51	31.31
March	29.50	5.10	23.38	19.22	21.92	16.79	32.05	31.47
April	29.51	6.19	26.47	19.65	23.91	20.92	32.95	32.09
May	30.27	9.25	29.41	20.86	24.90	24.25	34.14	33.05
June	32.94	14.75	30.49	29.51	29.99	28.53	35.01	33.37
July	34.52	19.19	30.52	32.68	33.74	29.05	35.51	33.52
August	35.15	20.97	30.55	34.38	35.27	29.46	35.73	33.75
September	34.46	18.53	30.64	31.65	33.69	30.46	35.80	33.55
October	31.42	15.71	30.71	21.97	27.89	31.15	35.73	33.97
November	27.19	9.80	30.68	13.17	21.83	28.95	34.43	33.44
December	28.68	9.50	28.42	17.16	23.30	23.17	32.10	32.64
Average	30.66	11.58	28.39	22.54	26.29	24.65	34.05	32.86
Range	8.25	16.16	7.33	21.21	16.35	15.28	3.75	2.67

The average value and the range of values are also indicated.

Table 5. Mean monthly modelled salinity for eight analysed estuaries calculated at the ocean-estuary defined cross-sections.

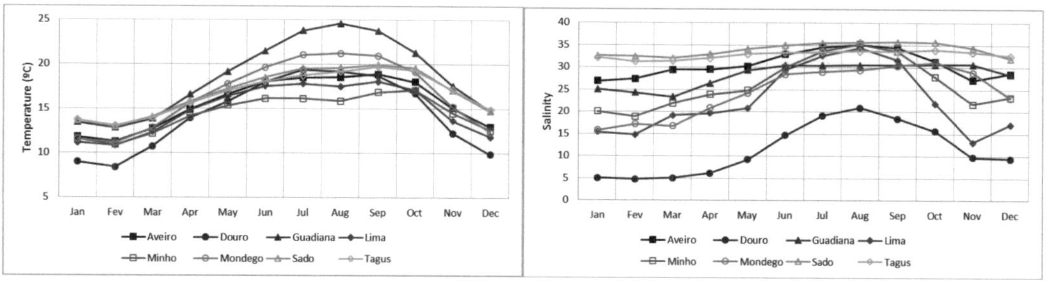

Figure 5. Mean monthly temperature (left) and salinity (right) obtained by numerical modelling at the estuarine mouth for eight Portuguese estuaries for the period 2010–2016.

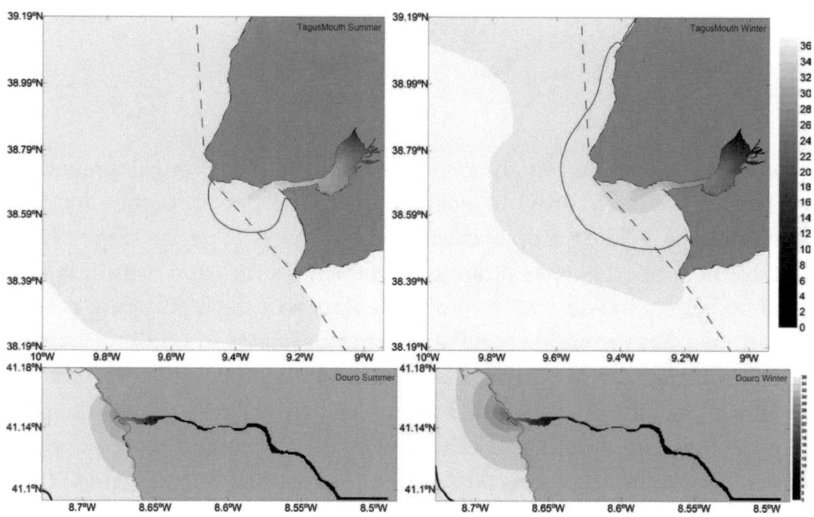

Figure 6. Mean summer (JJA) salinity (left) and winter (DJF) salinity (right) for the Tagus (top) and Douro (bottom) estuaries for the period 2010–2016. The continuous line indicates the salinity value of 35.75. Dashed line in Tagus graphs indicates the baseline for defining internal waters.

still far from what is considered as saline. Still, the highest ranges of salinities at their mouths are observed in the Lima estuary followed by the Douro estuary and far from the very stable Sado and Tagus estuaries.

The salinity differences between the estuarine fluxes and the ocean saline waters are responsible for generating fronts of different intensity that can influence the surface circulation of large coastal areas. In the Northern halve of Portugal, the high frequency of occurrence of rivers and estuaries is responsible for modifying local and coastal circulation contributing to the creation of haline fronts. As a result, a low salinity water lens, upper salinity limit below 35.7–35.8, denominated as Western Iberia Buoyant Plume (WIBP), extends along the northern Portuguese coast, being referred as a permanent feature of this part of the coast and with a varying intensity along the year [31, 32].

To illustrate the influence of the estuarine plumes on the near-shore ocean environment, the salinity average for the Douro and the Tagus estuary, the two most extreme estuaries of the study in terms of saline concentration, was obtained for the winter (DJF) and summer (JJA) seasons for the period 2010–2016. Both systems are characterised by large plumes in both seasons with salinity values below the WIBP upper limit (37.75) in the nearshore marine environment (**Figure 6**). The baseline in the Douro region that serves to identify the interior waters is out of the study domain. On the other hand, the Tagus plume extends further than the 1 nautical mile limit from baseline, the limit of the application of the MSFD and WFD, even during summer conditions. In the case of the Minho, Lima, Aveiro and Guadiana estuaries, the adopted baseline for the WFD and MSFD limit coincides with the coastline. For this reason, it is presumed that the estuarine plumes will have an effect beyond 1 nautical mile. This finding has implications for the implementation and evaluation of both EU Directives Research on estuarine plumes extension and their influence in the neighbouring waters is therefore essential.

5. Conclusions

Monitoring networks along the estuarine continuum from the river catchment into the open ocean should be encouraged in order to evaluate the transfer of properties and momentum in the land-ocean interface. While the open ocean and hydrometric monitoring networks are relatively well established, operational estuarine monitoring is far from consolidated. Numerical models are able to help in the design and implementation of the monitoring networks, validation of observed data and provide spatial and temporal data. In addition, numerical model completes the *in situ* data by providing non-observed variables and forecasts. Furthermore, modelling results allow for the calculation of complex parameters such as tidal prims, the area of influence of the estuarine waters and estuarine fluxes that would serve as boundary conditions for ocean regional models such as the offline coupling method applied to PCOMS [6]. This type of analysis provides valuable information to characterise the estuarine systems, as this was able to show that two neighbouring estuaries can differ largely in flow and salinity fluxes.

Acknowledgements

MARETEC acknowledges the ERDF funds of the Competitiveness Factors Operational Programme (COMPETE), and national funds from the Foundation for Science and Technology (FCT) (project UID/EEA/50009/2013). Part of this research was supported by the EMoSEM project, a two-year project (2013–2014) funded by the French National Research Agency (ANR) and the Belgian Science Policy (BELSPO) in the frame of EU FP7 ERA-NET Seas-era. River flow data were provided by the Portuguese National Institute of Water (INAG) through their portal SNIRH (http://www.snirh.pt). The authors are grateful to the water utility company SIMTEJO for providing the Algés coastal buoy data. The authors would like to acknowledge also the contribution of I. Ascione Kenov for the original development of the Douro estuary model application and to the IST meteorological group, nowadays part of MARETEC, and in particular to its current operator Jorge Palma for the maintenance of the operational models and delivery of the MM5 and WRF model results.

Author details

Francisco J. Campuzano[1*], Manuela Juliano[2], João Sobrinho[1], Hilda de Pablo[1], David Brito[3], Rodrigo Fernandes[3] and Ramiro Neves[1]

*Address all correspondence to: campuzanofj.maretec@tecnico.ulisboa.pt

1 MARETEC, Instituto Superior Técnico, Universidade de Lisboa, Lisbon, Portugal

2 LAMTec-ID, Universidade dos Açores, Praia da Vitoria, Portugal

3 Action Modulers, Consulting and Technology, Mafra, Portugal

References

[1] Pritchard DW. What is an estuary: physical viewpoint. In: Estuaries Lauf, GH, editor, American Association for the Advancement of Science Publication. No. 83. Washington, DC; 1967. pp. 3-5

[2] European Commission 2012. Links between the Marine Strategy Framework Directive (MSFD 2008/56/EC) and the Nature Directives (Birds Directive 2009/147/EEC (BD) and Habitats Directive 92/43/EEC (HD)). Available from: [Accessed: Sep 3, 2017]

[3] Borja A, Elliott M, Carstensen J, Heiskanen A-S, van de Bund W. Marine management – Towards an integrated implementation of the European marine strategy framework and the water framework directives. Marine Pollution Bulletin. 2010;**60**(12):2175-2186. DOI: 10.1016/j.marpolbul.2010.09.026

[4] Garel E, Nunes S, Neto JM, Fernandes R, Neves R, Marques JC, Ferreira Ó. The autonomous Simpatico system for real-time continuous water-quality and current velocity monitoring: Examples of application in three Portuguese estuaries. Geo-Marine Letters. 2009;**29**:331-341. DOI: 10.1007/s00367-009-0147-5

[5] Garel E, Ferreira Ó. Multi-year high-frequency physical and environmental observations at the Guadiana estuary. Earth System Science Data. 2015;**7**(2):299-309

[6] Campuzano F, Brito D, Juliano M, Fernandes R, de Pablo H, Neves R. Coupling watersheds, estuaries and regional ocean through numerical modelling for western Iberia: A novel methodology. Ocean Dynamics. 2016;**66**(12):1745-1756. DOI: 10.1007/s10236-016-1005-4

[7] AEMET IM. Atlas climático ibérico/Iberian climate atlas. Agencia Estatal de Meteorología, Ministerio de Medio Ambiente y Rural y Marino, Madrid. Instituto de Meteorologia de: Portugal; 2011. 80 pp. Available from: http://www.ipma.pt/resources.www/docs_pontuais/ocorrencias/2011/atlas_clima_iberico.pdf. [Accessed: Sep 3, 2017]

[8] Mishra AK, Coulibaly P. Developments in hydrometric network design: A review. Reviews of Geophysics. 2009;**47**(2):RG2001

[9] Neves R. The MOHID concept. In: Mateus M, Neves R, editors. Ocean Modelling for Coastal Management – Case Studies with MOHID. IST Press, Lisbon: Portugal; 2013. pp. 1-11

[10] Mateus M, Riflet G, Chambel P, Fernandes L, Fernandes R, Juliano M, Campuzano F, de Pablo H, Neves R. An operational model for the west Iberian coast: Products and services. Ocean Science. 2012;**8**:713-732

[11] Campuzano FJ, Fernandes R, Leitão PC, Viegas C, de Pablo H, Neves R. Implementing local operational models based on an offline downscaling technique: The Tagus estuary case. In: 2.as Jornadas de Engenharia Hidrográfica. Lisbon: Portugal. Extended abstracts; June 20-22, 2012. pp. 105-108

[12] Brito D, Campuzano FJ, Sobrinho J, Fernandes R, Neves R. Integrating operational watershed and coastal models for the Iberian Coast: Watershed model implementation – A first approach. Estuarine, Coastal and Shelf Science. 2015;**167**(Part A):138-146. DOI: 10.1016/j.ecss.2015.10.022

[13] Skamarock WC, Klemp JB, Dudhia J, Gill DO, Barker DM, Wang W, Powers JG. A description of the advanced research wrf Version 2. Tech. Rep., National Center for Atmospheric Research. 2005

[14] Grell GA, Dudhia J, Stauffer D. A description of the fifth-generation Penn State/NCAR Mesoscale Model (MM5). NCAR Technical Note NCAR/TN-398+STR. 1994

[15] Trancoso AR. Operational Modelling as a Tool in Wind Power Forecasts and Meteorological Warnings. [PhD thesis]. Technical University of Lisbon: Instituto Superior Técnico; 2012

[16] Kenov IA, Campuzano F, Franz G, Fernandes R, Viegas C, Sobrinho J, de Pablo H, Amaral A, Pinto L, Mateus M, Neves R. Advances in modeling of water quality in estuaries. In: Finkl CW, Makowski C, editors. Remote Sensing and Modeling. Cham (ZG) Switzerland: Springer International Publishing AG; 2014. pp. 237-276. DOI: 10.1007/978-3-319-06326-3_10

[17] Saraiva S. Modelação ecológica da Ria de Aveiro: o papel das macroalgas [MSc thesis]. Portugal: Technical University of Lisbon; 2005

[18] Sobrinho JL. Balanço de nutrientes na plataforma continental ao largo da Região de Aveiro. [MSc thesis]. Universidade de Lisboa, Portugal: Instituto Superior Técnico; 2012

[19] Saraiva S, Pina P, Martins F, Santos M, Braunschweig F, Neves R. Modelling the influence of nutrient loads on Portuguese estuaries. Hydrobiologia. 2007;**587**(1):5-18

[20] Kenov IA, Garcia AC, Neves R. Residence time of water in the Mondego estuary (Portugal). Estuarine, Coastal and Shelf Science. 2012;**106**:13-22

[21] Martins F, Leitão P, Neves R. Simulating vertical water mixing in homogeneous estuaries: The SADO estuary case. Hydrobiologia. 2002;**475**(476):221-227

[22] Fernandes L. Modelling of arsenic dynamics in the tagus estuary. [MSc thesis]. Portugal: Technical University of Lisbon; 2005

[23] Hume TM. Tidal prism. In: Schwartz ML, editor. Encyclopedia of Coastal Science. Netherlands: Springer; 2005. pp. 981-982. DOI: 10.1007/1-4020-3880-1_320

[24] Picado A, Dias JM, Fortunato A. Tidal changes in estuarine systems induced by local geomorphologic modifications. Continental Shelf Research. 2010;**30**:1854-1864. DOI: 10.1016/j.csr.2010.08.012

[25] Vicente CM. Caracterização hidráulica e aluvionar da Ria de Aveiro, Utilização de modelos hidráulicos no estudo de problemas da Ria. In: da Ria de Aveiro III J, editor. Edição da Chara Municipal de Aveiro, Aveiro, Portugal, 41-58, 1985

[26] Ferreira JG, Simas T, Nobre A, Silva MC, Schifferegger K, Lencart-Silva J. Identification of sensitive areas and vulnerable zones in transitional and coastal Portuguese systems. Application of the United States National Estuarine Eutrophication Assessment to the Minho, Lima, Douro, Ria de Aveiro, Mondego, Tagus, Sado, Mira, Ria Formosa and Guadiana systems. Lisbon: INAG - Instituto da Água and IMAR – Institute of Marine Research; 2003. 168 p

[27] Verhagen HJ. Hydro-morphogical study Douro estuary: Part 3 – Tidal current calculations. Report P613 for Administraçâo dos Portos do Douro e Leixões (ADPL). Delft: Hydrodynamic; 1982. 25 p

[28] Duarte AALS, Vieira JMR. Effect of tidal regime on estuarine residence time spatial variation. In: Mastorakis N, editor. Proceedings of the 5th International Conference on Energy, Environment, Ecosystems and Sustainable Development (EEESD '09). 28-30 September 2009, Athens (Greece) pp. 240-245

[29] Brito P, Costas S, Rebêlo L, FitzGerald D. Record of ebb-channel shifting within an ebb-tidal delta, the Sado Estuary, Portugal [Abstract]. In: International Coastal Symposium 2013: ICS2013: Book of Abstracts. Plymouth, UK. 8-12 April 2013. 33 p

[30] Neves FJ. Dynamics and hydrology of the Tagus estuary: Results from in situ observations [PhD thesis]. Portugal: Universidade de Lisboa; 2010

[31] Peliz Á, Rosa TL, Santos AMP, Pissarra JL. Fronts, jets, and counter-flows in the western Iberian upwelling system. Journal of Marine Systems. 2002;**35**(1-2):61-77

[32] Peliz Á, Dubert J, Santos AMP, Oliveira PB, Le Cann B. Winter upper ocean circulation in the western Iberian Basin—Fronts, eddies and Poleward flows: An overview. Deep-Sea Research I. 2005;**52**:621-646

Chapter 2

Weather, Hydrological and Oceanographic Conditions of the Northern Coast of the Río de la Plata Estuary during ENSO 2009–2010

Ernesto Brugnoli, José Verocai, Pablo Muniz and Felipe García-Rodríguez

Additional information is available at the end of the chapter

http://dx.doi.org/10.5772/intechopen.71808

Abstract

Climatic, hydrological, and oceanographic conditions were determined during the 2009–2010 El Nino/Southern Oscillation (ENSO) on the north coast of the Río de la Plata (RdlP) estuary. The maximum monthly rainfall was observed in the middle and upper La Plata basin during September 2009 and February 2010 ("El Niño" phase, (EN)). The monthly flow of RdlP showed an increase with rainfall and significant differences between ENSO phases. The wind stress showed fluctuations in both phases, being less intense during EN, during which time maximum flow of RdlP was observed. During the EN phase, increased precipitation contributed to variations in salinity and absence of water column stratification in the north coast of RdlP. This was also associated with variations in Secchi depth, oxygen saturation, and nutrient concentrations. The spatial location of the turbidity front was associated with the flow of the RdlP and wind stress, thus conditioning the physicochemical characteristics of the water column, mainly during EN phase.

Keywords: estuarine dynamics, river flow, physicochemical properties, ENSO, Río de la Plata

1. Introduction

"El Niño/Southern Oscillation" (ENSO) is a global phenomenon related to the ocean-atmosphere interactions in the Equatorial Pacific, with cycles consisting in a warm phase ("El Niño"), periods with neutral years ("Neutral"), and a cold phase ("La Niña"), with a 3–7 years periodicity [1, 2]. This event activates changes in the general atmospheric circulation, causing climate and hydrologic changes in continental areas, coastal areas, and tropical and extra-tropical zones of the Pacific or Atlantic Oceans [3, 4]. These modifications cause changes in the systems

INTECH
open science | open minds

© 2018 The Author(s). Licensee InTech. This chapter is distributed under the terms of the Creative Commons Attribution License (http://creativecommons.org/licenses/by/3.0), which permits unrestricted use, distribution, and reproduction in any medium, provided the original work is properly cited.

physicochemical conditions at different organizational levels, impacting on the composition and abundance of benthic, planktonic, and nektonic communities [5–7].

In the La Plata basin, seasonality of the climatic elements determining the hydrologic cycles is ruled by the South Atlantic anticyclone, which is more intense in winter. The region has a warm season (October–April) with an average rainfall of 5.5 mm d^{-1} and maximum values near 9 mm d^{-1} and a cold season (May–September) with average rainfall less than 2 mm d^{-1} [8]. The alteration of the rainfall regime is one of the strongest signals of the changes caused by the ENSO event for the southern region of South America. During the warm phase, there is an increase in the spring rains, while during the cold phase, there is a decrease in rainfall [9, 10]. On the other hand, flows of the hydrological systems of the La Plata basin, including those of the main tributary rivers of the RdlP estuary (Parana and Uruguay rivers), are highly sensitive to modifications in the rainfall regime, with inter-annual and inter-decadal variability, being affected by ENSO events in their warm phase [11, 12].

Surface ocean temperature is the most analyzed oceanographic variable regarding the "El Niño" phase and its ecosystem effects in oceanic and coastal marine zones [13]. Nevertheless, estuarine zones are expected to experience a greater impact of hydrologic effects (fresh water or ocean water inputs) [14]. ENSO impacts in coastal zones include increased rainfall, hydrological changes in basins, modifications in river flow [15], and generation of inter-annual variations in fresh water discharges for estuarine systems [16, 17].

ENSO is considered in the RdlP as a large-scale atmospheric forcing effect on the discharges of the tributary rivers [15, 17], changes in salinity and nutrients [18], as well as in the location and location of turbidity or salinity fronts [19, 20]. Several studies have described the relationship between ENSO events, RdlP flow, and salinity of the estuarine system [18, 19, 21]. According to [16], surface sediments on the north coast of RdlP showed a variability in the trophic state related to the ENSO event, with differences between the phases ("El Niño" and "La Niña-Neutral").

The aim of this study was to identify the main environmental drivers that promote modifications in the oceanographic conditions on the north coast of the RdlP, during ENSO 2009–2010. We measured the meteorological (rainfall and wind), hydrological (principal rivers flow), and oceanographic (temperature, salinity, oxygen saturation, Secchi depth, chlorophyll *a*, and total nutrients) conditions during both ENSO phases in the north coast RdlP (Montevideo coastal zone).

2. Methodology

2.1. Study area

The La Plata basin is the second largest river basin in the South America, covering an area of 3.1 $\times 10^6$ km^2; it includes five countries (Argentina, Brazil, Bolivia, Paraguay and Uruguay) with some of the most populated cities in South America including Sao Paulo, Buenos Aires, and Montevideo. The main rivers draining into the estuary of RdlP are the Parana and Uruguay rivers (**Figure 1A**).

The RdlP is an estuarine system characterized by the presence of a salt wedge, low river discharge seasonality, low tidal amplitude (< 1 m), an extensive and permanent connection to the sea, and high susceptibility to atmospheric drivers because of its large size and its shallow depth. The mean annual RdlP flow governs the salinity and has monthly to inter-annual variations of 25,000 m^3s^{-1} [22–24]. In this system, several authors report biological processes [18], impacts on water quality, or presence of invasive alien species [20] associated with the spatiotemporal fluctuation of its front area [25].

The montevideo coastal zone (MCZ) is located on the north coast and middle zone of the RdlP between the mouths of the Santa Lucia River and Carrasco stream with an approximate extension of 50 km (**Figure 1B**). The largest city in Uruguay, Montevideo, covers ≈ 49% of the coastal zone [26]. The MCZ is characterized by high levels of industrial and harbor activities [26, 27] but also with urban spaces for recreation (sandy beaches and rocky shores), diving, artisanal or sport fishing areas, and conservation areas (protected area Punta Yeguas and Santa Lucía Wetlands). Several studies in the MCZ identified a contamination gradient from the innermost Montevideo Bay area to the outer and adjacent coastal zone. Three contamination areas have been identified within the region: high (internal Montevideo Bay and Montevideo Harbor), medium (external Montevideo Bay and adjacent coast), and low impact (adjacent coastal zone, includes Punta Brava and Punta Yeguas) regions [27]. Montevideo Bay is 10 km^2 and has a mean depth of ≈ 5m. In this system, nutrients (nitrogen, phosphorus), heavy metals, and hydrocarbons pollution have been associated with significant degradation of ecosystem [27–31]. These pollutants are derived from industrial activities, from "La Teja" refinery, harbor operations (i.e., navigation, bulk loading and dredging), or contributions from domestic and

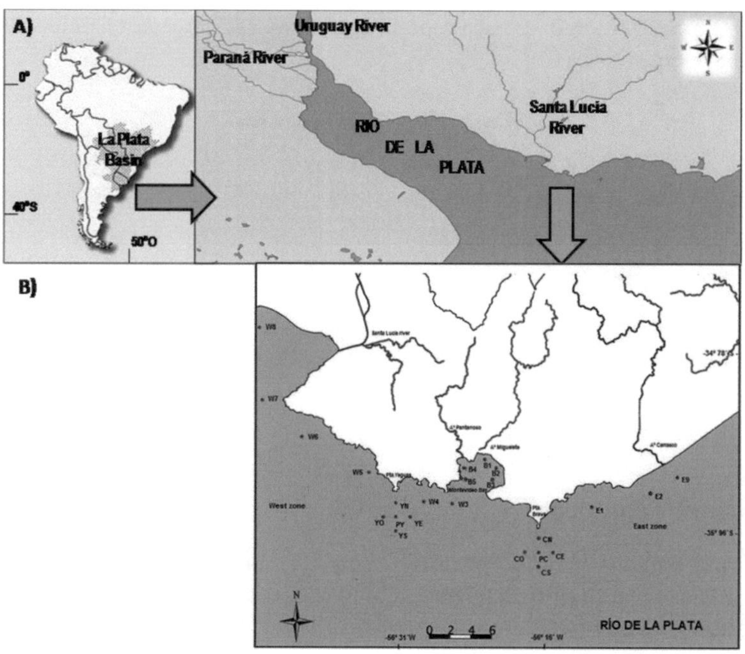

Figure 1. Location of the La Plata Basin, RdlP and Montevideo coastal zone (MCZ) with Punta Brava and sampling sites.

industrial effluents from Pantanoso, Miguelete, and Seco streams. In the East zone, 2000 m off the coastline (Punta Brava), the submarine outflow of the sanitation system of Montevideo is located. In addition, in this zone, there are recreation areas with presence of sandy beaches, alternating with rocky shores. In the West zone (Punta Yeguas), another submarine outflow with similar characteristics in Punta Brava will be installed soon.

2.2. Weather conditions

Variability and magnitude of ENSO were determined by the Ocean Niño Index (ONI) considering the monthly anomalies of ocean surface temperature (SST) in the Niño 3.4 region (2009–2011); values were obtained from http://www.cpc.ncep.noaa.gov/products/analysis_monitoring/ensostuff/ensoyears.shtml.

Total monthly precipitation to the upper zone of La Plata basin (22°00′–40°00′ S and 48°00′–64°00′ W) was obtained from the Global Precipitation Climate Center (GPCC; http://gpcc.dwd.de). The product of monitoring of precipitation on earth surface was analyzed (1.0° × 1.0°). Values are expressed as rainfall "isoareas" (mm $month^{-1}$), using monthly values (March 2009 and August 2011).

Wind speed and direction were collected on an hourly basis at the Punta Brava meteorological station, RdlP (34°56′S and 56°09′W; March 2009 and August 2011). We calculated the average monthly speed (± SD), minimum, and maximum and are expressed in m s^{-1}. We did not consider maximum wind speeds (streaks of wind) in these calculations. To determine the direction of the prevailing winds, 16 quadrants were considered, and monthly relative frequency was calculated. Wind stress (Eq. **(1)**) was calculated as surface force of the wind on the water column; stress values of positive or negative wind were determined by the average monthly wind direction.

$$\tau(\text{Pa}) = \beta a * \text{CD} * (V\ \text{wind})^2 \tag{1}$$

where βa = air density, 1.2 kg m^{-3}, CD: drag coefficient: 0.0013, and V wind: wind speed.

2.3. Hydrological conditions

The daily river flow of the Uruguay and Parana rivers was obtained from the National Water Institute (www.ina.gov.ar) for the period 2009–2011. The river flow of the RdlP was obtained by adding the daily river flow of the Uruguay and Parana rivers. Monthly averages (± SD) were calculated and expressed as $m^3\ s^{-1}$.

2.4. Oceanographic conditions

Eighteen oceanographic surveys comprising 25 stations (located at 2000 m of coastline, depth 3–10 m; **Figure 1B**) were conducted in the MCZ between March 2009 and July 2011. The study area was divided into the Eastern zones (E1, E2, E9, PC, CN, CE, CS, CW stations), Western zones (W3 to W8, PY, YN, YE, YS, YW stations), and Montevideo Bay (MB) (B1 to B5 stations).

Water temperature and salinity were determined *in situ* at surface and bottom waters using a multiparameter *YSI Pro plus*. During 11 surveys, oxygen saturation was determined with multiparameter in surface waters and water transparency calculated using a Secchi disk (30 cm diameter). In addition, surface water samples were collected using a Kemmerer (2L) bottle to determinate chlorophyll *a* (Chl *a*) and nutrients (total nitrogen and total phosphorous: TN, TP) concentrations.

2.5. Laboratory analysis

Chlorophyll *a* concentration was quantified by spectrophotometric analysis using GF/F filters extracted in 90% acetone [32]. Determination of TN and TP was performed according to [32, 33] with previous digestion according to [34].

2.6. Data analysis

For the analysis of the temporal variation associated with ENSO, the EN and "La Niña-Neutral" months were defined according to Oceanic Niño Index (ONI). Nonparametric analyses (U Mann-Whitney) were performed to assess temporal differences (ENSO phases) in hydrological variables (Uruguay, Parana, and RdlP flow), and spatiotemporal differences in oceanographic parameters (temperature and salinity). Nonparametric correlations ($R_{s,}$ Spearman) were performed between climatic (wind stress, Niño indexes), physicochemical parameters, and river flow (RdlP, Uruguay, and Paraná). Linear relationships between river flows and average salinity were performed; both variables were transformed by log (x + 1) to fit the assumption of normality. A principal component analysis (PCA) was performed with all the physicochemical parameters (temperature, salinity, oxygen saturation, Secchi depth, Chl *a*, TN, and TP). The variables were log-transformed (x + 1), standardized, and Varimax type rotation was considered. We considered 99% and 95% significance levels for the different analyses. The statistical analyses were performed with the SPSS and CANOCO program [35].

3. Results and discussion

3.1. Weather conditions

The ONI showed the minimum values during October–December 2010 (<−1.5°C) and maximum values in December 2009 and January 2010 (<1.5°C; **Figure 2**). The 2009–2010 ONI values allowed for the identification and development of an ENSO event with an "El Niño" (EN) phase (July 2009–April 2010), a subsequent "La Niña" (LN) cold phase (June 2010–April 2011; September–December 2011), and neutral conditions (April–June 2009, April–May 2010, and May–August 2011) [2, 36].

The 2009–2010 ENSO event was classified as "moderate to strong" [37]. When comparing the maximum 2009–2010 value of this indicator (October 2009–January 2010: 1.2) with those of recent ENSO events, we found that they were smaller than those corresponding to strong events (October–January: 1982–1983: 2.1; 1997–1998: 2.3; 2015–2016: 2.2), although they were

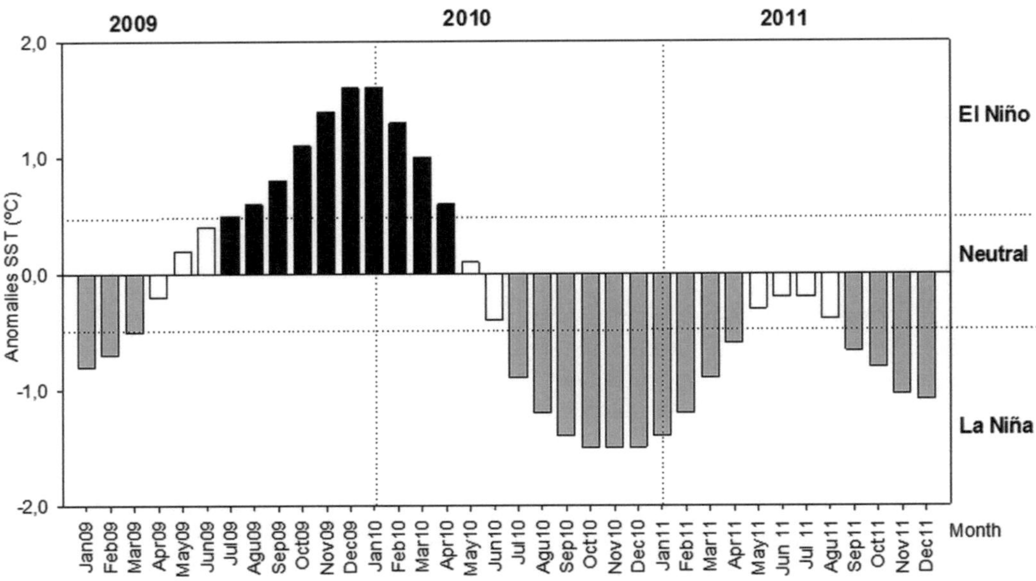

Figure 2. ONI index represented by SST anomalies (Niño 3.4 region) and ENSO 2009–2010 phases. Horizontal lines >0.5°C and <−0.5.

similar to those of moderate events (1986–1987 and 1991–1992) [37]. Although in the range of previous moderate events, the 2009–2010 ENSO had characteristics of its own, including high SST anomalies in the Central Pacific and the fastest reported transition to the LN phase [2, 36]. The warm phase of the 2009–2010 was classified into the "WP" (warm pool) type and differs from the " Eastern Pacific" (EP) type due to the fact that SST anomalies happen in the central and not in the western zone of the Pacific Ocean [2]. It is also known as CP or WP, dateline ENSO or "El Niño Modoki" [2], and has different teleconnections and differential climatic impacts compared to the "EP" [38]. The analysis of the effects in the MCZ of the 2009–2010 ENSO, with "El Niño Modoki" characteristics, will permit the identification of variations in the behavior of meteorological variables and their impacts in the hydrologic behavior of the main hydrological systems of the La Plata basin, as well as over the RdlP estuary and its oceanographic conditions.

Total monthly precipitation for the La Plata basin during the period March 2009–August 2011 showed minimum values in April 2009 (0–10 mm month^{-1}) and maximum during September 2009 and February 2010 (**Figure 3**). The highest rainfall for the upper basin was reported in November 2009 (600 mm month^{-1}), while the values in February 2010 exceeded 550 mm month^{-1}. The teleconnections analyzed for the ENSO event in South America (Atlantic coast), specifically for the La Plata basin, are mainly related to their effect on rainfall variability [12]. This variability caused anomalies in surface air temperature in Argentina and southern Brazil during the "El Niño Modoki" (June–September 1979–2004), as well as higher-than-average rainfall between the months of December–February (1979–2004 period) [38]. In addition, [10] found an increase in spring rainfall during the EN warm phase. In this study, we found that the maximum rainfall values in the middle and upper basin of the Uruguay River were recorded

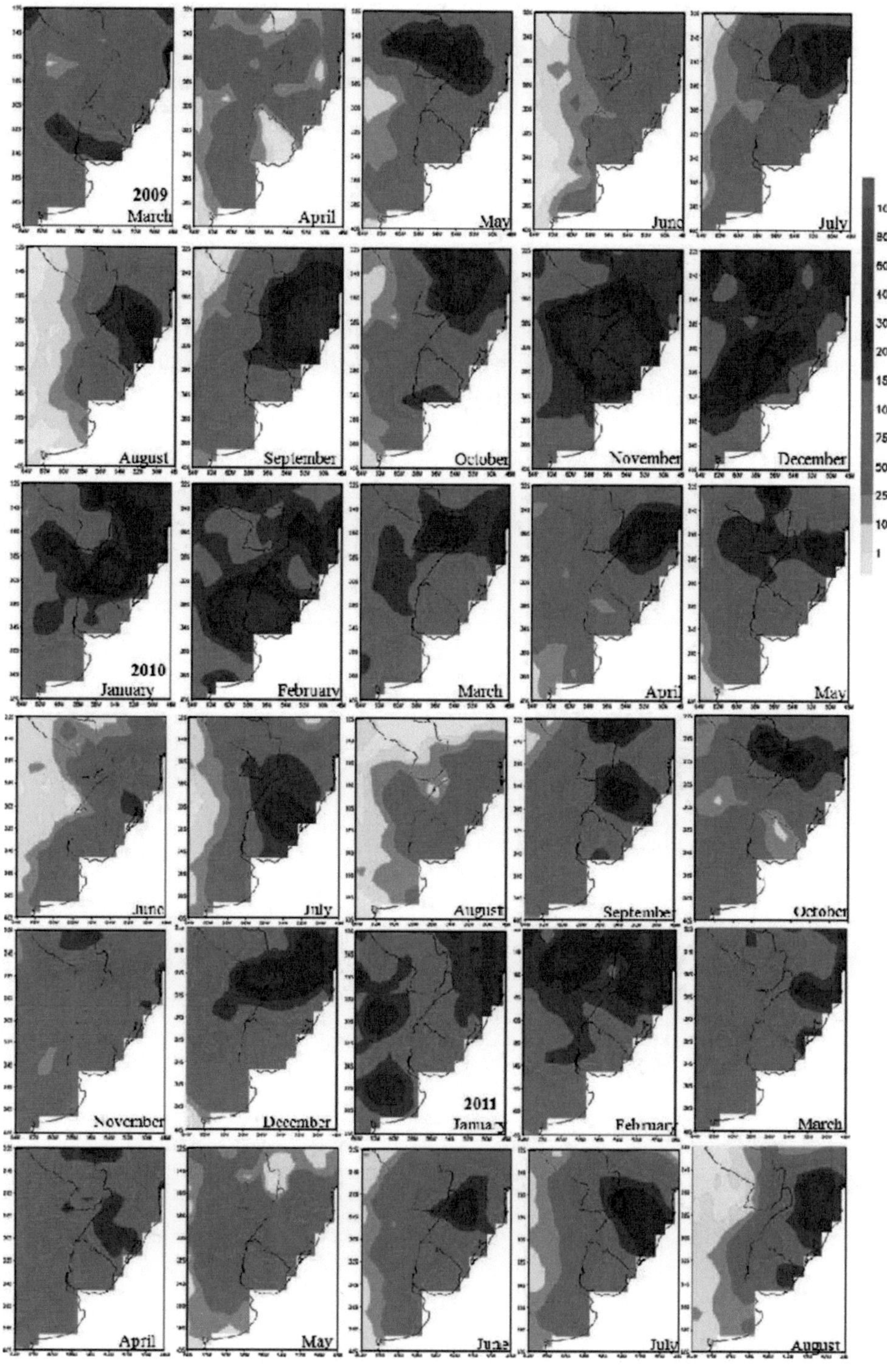

Figure 3. Total monthly precipitation (mm month^{-1}) on the surface of the earth to the region 22°00′–40°00′ S and 48°00′–64°00′ W.

in November 2009 and December 2010, corresponding to the warm ENSO phase (EN: July 2009–April 2010) and to the spring months of the calendar year (September–December) and minimum rainfall values (April 2009 and October 2010) which corresponded to those months of Neutral and LN phase.

The average monthly wind speed between March 2009 and August 2011 was 5.8 m s^{-1} ± 2.9 (n = 6722) (2009: 6.1 ± 3.3; 2010: 5.6 ± 2.9; 2011: 5.9 ± 2.7); wind speed oscillated between 0 m s^{-1} (stills) and a maximum of 23.7 m s^{-1} in July 2009 (**Figure 4A**). The most frequent wind direction was the first quadrant (N-E) with similar annual rates of occurrence (2009: 42%, 2010: 43%, 2011: 44%); Easterly winds were the most frequent (2009: 11%, 2010: 16%, 2011: 14%; **Figure 5**). Wind stress showed minimum values during months of the EN phase (minimum September 2009:−0.076 Pa, ESE) with a peak during June–July 2009 (0.068 Pa direction N and O; **Figure 4B**). During ENSO events in the RdlP, [18] found no variations in wind speed and reported the predominance of ESE and NE winds. In addition, [20] found that during EN years, there is an increase of ESE to SE winds. In this study, we observed that in EN months (September 2009–April 2010), there was an increase in the monthly frequencies of winds in eastern direction.

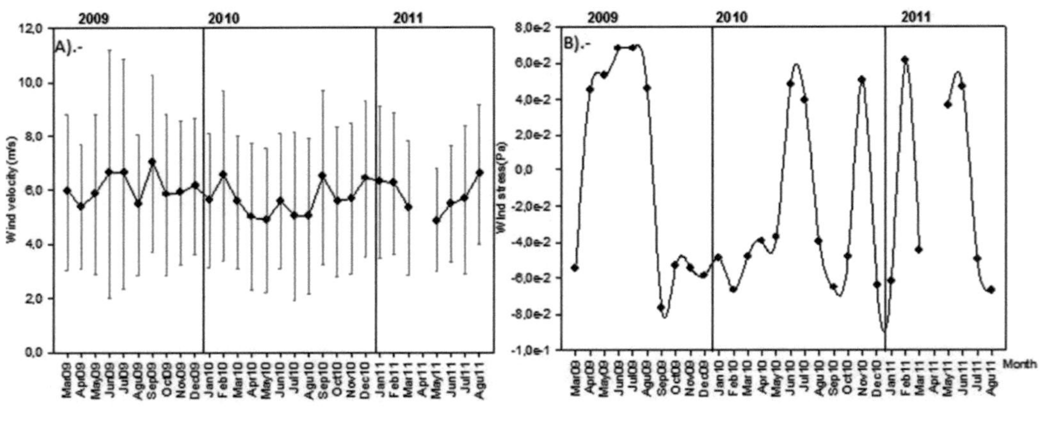

Figure 4. Average monthly wind speed (± SD) (m s^{-1}) (A) and wind stress (Pa) fluctuations (B) between 2009 and 2011.

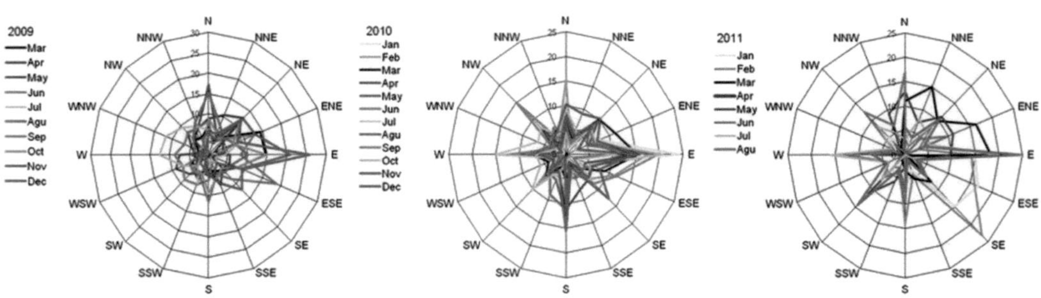

Figure 5. Monthly absolute frequencies of wind direction between 2009 and 2011 (2009: March to December, 2011: January–August).

3.2. Hydrological conditions

The average monthly flow of the RdlP, Uruguay, and Parana rivers showed an oscillation between minimum values during January and June 2009 and maximum between December 2009 and March 2010. Differences in the months of maximums discharges of the Uruguay and RdlP (December 2009) with the Parana River (March 2010) were observed (**Figure 6**).

The great rivers of the La Plata basin and tributary rivers of the RdlP are highly sensitive to rainfall variations, their flows being impacted during EN phases [11, 12]. In this study, we found strong correlations in the hydrological behavior of the Uruguay and Parana rivers in response to the increase of rainfall during the 2009 spring, coinciding with the EN warm phase. The maximum flow of the Uruguay and Parana rivers was recorded during December 2009–March 2010, and the minimum flow corresponded to those months prior to the development of the warm phase (January and July 2009). The RdlP flow had two peaks: the first in December 2009, coinciding with the maximum observed flow for the Uruguay River, and a second during the months of February–March 2010, coinciding with the maximum flow of the Parana River, previously indicated for this system by [15].

Monthly average flow of RdlP, Uruguay, and Parana rivers showed significant differences between the two ENSO phases (Mann-Whitney U: RdlP: $Z = -12.17$; Uruguay River: $Z = -11.09$; Parana River: $Z = -8.15$; $P < 0.01$). Significant associations between the flow rates of the three water systems were observed during both phases: EN: RdlP-Uruguay River (R_s $(_{0.025,\ 10}) = 0.697$) and RdlP-Parana River (R_s $(_{0.04,\ 10}) = 0.818$), "La Niña-Neutral": RdlP-Uruguay River (R_s $(_{0.00,\ 21}) = 0.823$) and RdlP-Parana River (R_s $(_{0.00,\ 21}) = 0.970$). For the period December 2009 to March 2010, the RdlP flow exceeded 40,000 $m^3\ s^{-1}$, the greatest average flow registered in 16 years (1999–2014). Nevertheless, average values of the RdlP flow during the 2009–2010 ENSO event (24,806 $m^3\ s^{-1}$) and its different phases (EN: 32,933 $m^3\ s^{-1}$; "La Niña-Neutral," LNN: 22,039 $m^3\ s^{-1}$) were similar to

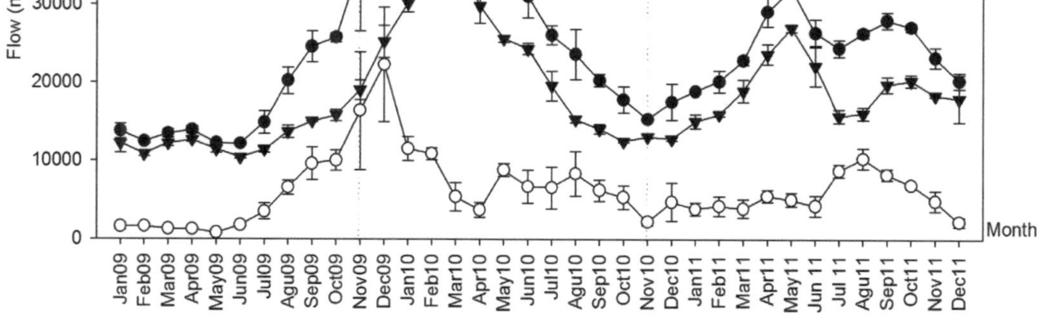

Figure 6. Monthly average flow (± SD) ($m^3\ s^{-1}$) of RdlP, Uruguay and Parana rivers between 2009 and 2011.

other ENSO events previously registered for the RdlP (1961, 2008: 24,700 $m^3 s^{-1}$ and 1998, 2008: 24,000 $m^3 s^{-1}$; [19, 20]). Average flow of the Uruguay River during low (3000–4000 $m^3 s^{-1}$) or high (< 7000 $m^3 s^{-1}$) discharge periods is lower than the ones found for LNN (Q Uruguay River = 5238 $m^3 s^{-1}$) and EN (Q Uruguay River = 9970 $m^3 s^{-1}$) in this study, although they fall into the range reported for ENSO events, characterized by their high variability (1000–20000 $m^3 s^{-1}$; [19, 20]). On the other hand, the average discharge of the Parana River for the 1884–1975 period (17,000 $m^3 s^{-1}$) ranged between 8000 and 22,000 $m^3 s^{-1}$ [39], falling within the range of the flow found in this study.

3.3. Oceanographic conditions

The average water temperature (surface-bottom) displayed a seasonal pattern with minimum temperatures in winter (July 2010: 10.4 ± 0.2°C and July 2011: 10.6 ± 0.6°C) and maximum in summer (January 2010: 25.9 ± 1.1°C; **Figure 7**). No significant differences between surface and bottom or between areas of the study were detected ($P > 0.05$). The average salinity ranged between minimum values during July 2010 (0.6 ± 0.5) and maximum values in March 2009 and January 2011 (21.2 ± 6.1, 22.3 ± 8.8, respectively). Salinity values in surface and bottom waters showed no significant differences for most of the study; however, differences were found at the beginning (March and July 2009, Mann-Whitney U: $Z = -3.46$ and $Z = -2.57$; $P < 0.01$) and at the end of the study period (June and July 2011; Mann-Whitney U: $Z = -2.95$ and $Z = -3.06$; $P < 0.01$). During the warm phase, the MCZ comprised brackish water with oligo to mesohaline conditions (maximum salinity: 15) and a temperature range between 15 and 28°C; in the cold-neutral phase, the water column salinity was highly variable (rank salinity: 0.1 to 33), with temperatures between 10 and 29°C. In July 2009, salinities were higher (15–30 range), although average water temperatures were lower than that during the EN phase (**Figure 7**).

Water temperature demonstrated greater temporal than spatial variation, with the lack of any differences between different depths or sampling study zones. Minimum and maximum values corresponded to winter and summer, respectively, and were associated with environmental

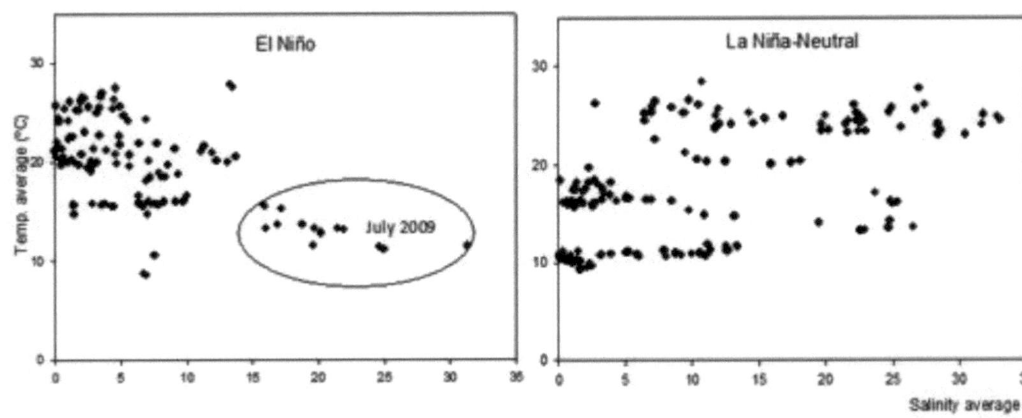

Figure 7. T-S diagrams (temperature and salinity average) during ENSO phases.

factors. Similar results were found in long time series in MCZ, and the seasonal patterns are consistent with previous studies for the middle RdlP [22, 24, 25, 39]. Reduced variation coefficients were found in summer (February 2010: 1%), which increased in July 2009 (16%). During this investigation, the average water temperature in July 2009 was 12.4 ± 2.1°C higher (10.7 ± 0.5°C) than during the other winter months study period (July 10, June and July 11). These results suggest an increase in temperature in the months prior to the development of EN over the MCZ. In this regard, [21] found a variability, related with years of pre- ENSO events in winter (April–October), in the anomalies of ocean surface air temperature (SST) in eight points of the South-West Atlantic region. The EN phase is characterized by negative anomalies in the SST in the Brazil Current, while during LN phase, cold anomalies were recorded in the Brazil Current and warm ones in the Malvinas Current. In the RdlP estuary, the influence of ocean bodies over adjacent coastal waters can be observed: sub-Antarctic cold waters between autumn and late spring and subtropical waters between late spring and autumn [40]. The anomalous records of July 2009 could be related to the LN phase prior to the effect of the EN event in the coastal zone of the RdlP.

Salinity is the main physicochemical variable of the water column, or "master parameter" of the RdlP, operating as regulator of biogeochemical, ecological, and sedimentological processes [18, 20, 22, 24, 41]. During most of the months of the EN (October 2009–July 2010) and LNN (September 2010–February 2011), we observed a mixed water column with no salinity stratification. Nevertheless, in the months prior to the EN (March and July 2009) and in LNN months (June and July 2011), the water column within MCZ was stratified. Vertical mixing of the water column in the study area may be generated by the predominant winds (speed and direction) [24] or by the fresh water inputs from the Uruguay and Parana rivers [18, 24, 40]. We identified the flow as the predominant driver, over wind stress, of the extension of the discharge plume of the RdlP during the EN months; this effect may also explain the lack of stratification along coast north of the RdlP during EN phase.

According to studies performed with long time series (1935–1975 Montevideo Bay; 1971–1991 Punta Brava), salinity in MCZ shows annual variations, with minimum monthly average values in autumn-winter and maximum ones in summer; minimum salinity values may occur throughout all the year, particularly in February, May, June, August, October, and December, although they have not been registered in January and rarely in July [40]. The minimum average salinity values were found during July 2010; however, according to [40], this month rarely displays such minimum salinity values. In addition, the mean salinity was 5.0 ± 4.6 higher during EN phase and even higher (12.7 ± 9.8) in LNN months. According to the inverse relationship between salinity and RdlP flow, the anomalous values are related to the ENSO 2009–2010 event, to the minimum salinity values during the EN months·(October 2009–July 2010) and to the maximum values during LNN, respectively, which are associated with maximum and minimum values of the RdlP flow.

The first two components of PCA accounted for 67.5% of the total variance of the spatiotemporal variability. Principal component 1 (38.6%) showed a positive correlation with salinity (0.736), Secchi depth (0.798), and oxygen saturation (0.589), interpreted as gradient of hydrological variability. Principal component 2 (28.9%) showed a positive correlation with temperature

(0.882) and Chl *a* (0.826), interpreted as gradient of seasonal variability. Associated with axis 1, stations of the EN phase with lower salinity and water transparency (Secchi depth) and higher nutrient (TN and TP) were located. The stations of the LNN phase are located to the right of axis 1 and are characterized by higher values of oxygen saturation, water transparency, and salinity. Associated with axis 2, stations with higher (upper) or lower (lower) temperature values and Chl *a* concentrations are observed (**Figure 8**).

The location and presence of the turbidity front is one of the environmental features regulating ecological processes in the RdlP [19, 24, 25]. The influence of ENSO events on this feature have been poorly studied [24]. Our results agree with those of [40], who found that the turbidity front was related to the salinity and discharge flow of the RdlP. In this study, we showed the relationship between the RdlP flow and the spatial location of the turbidity front that conditioned the physicochemical characteristics of the water column of the MCZ, mainly during the EN phase. In **Figure 9**, we show satellite images (MODIS) of the RdlP where different locations of the turbidity front are depicted. Before the development of the EN phase (February 2009), the turbidity front was located west of the MCZ. During the EN phase (July 2009–April 2010), the turbidity front was located adjacent to the MCZ, while during the LNN (October 2010–January 2011), the front was again west of the MCZ, with less turbid waters in the MCZ. The monthly average Secchi depth was directly correlated with salinity, thus reflecting the occurrence of low turbidity waters in the MCZ during the low discharge flows of the RdlP. In addition, the PCA showed an arrangement of the

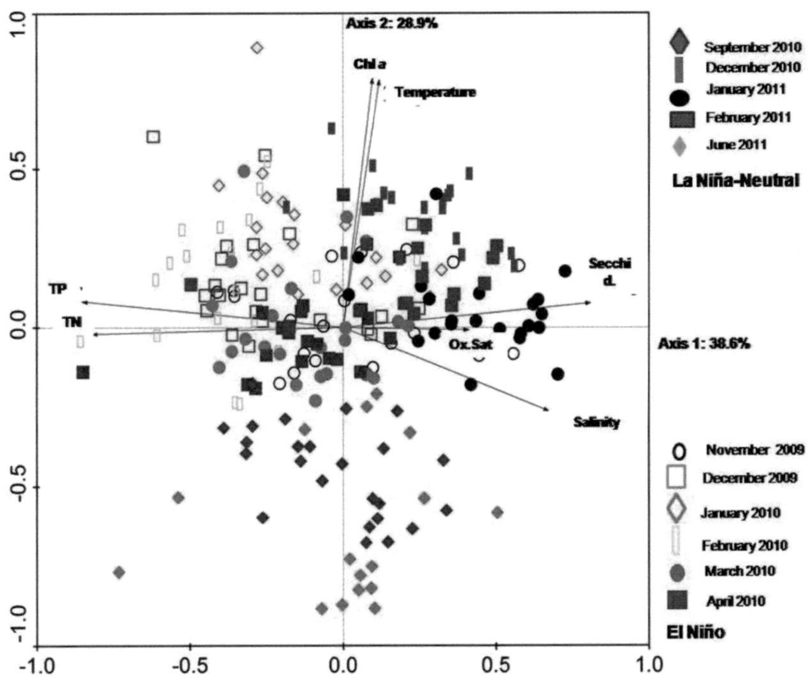

Figure 8. Principal component analysis diagram of physicochemical parameters classified according to ENSO phases ("El Niño" and "La Niña-Neutral"). TN (Total nitrogenous), TP (Total phosphorus), Ox. Sat. (oxygen saturation), Secchi d. (Secchi depth), Chl *a* (chlorophyll *a*).

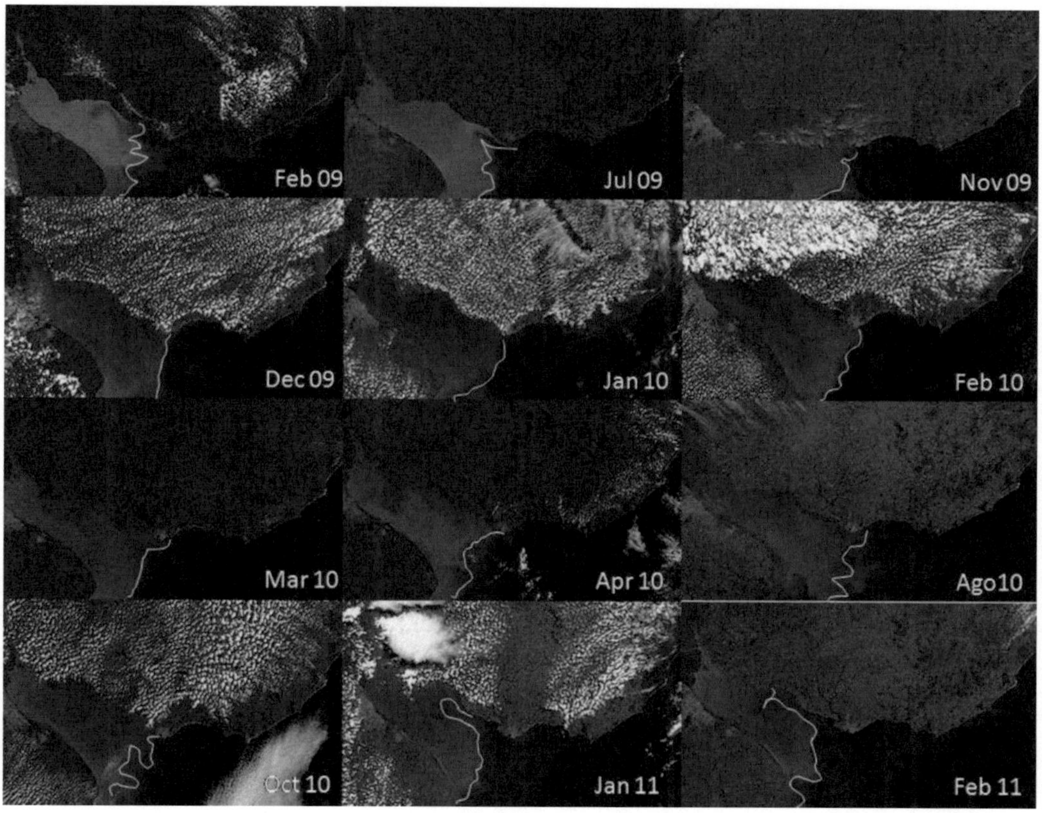

Figure 9. Turbidity front position (white line) during the study period. Images MODIS (AERONET-CEILAP-BA-Subset-Aqua-1 km-true-color). From: http://lance.modaps.eosdis.nasa.gov/imagery/subsets/?subset=AERONET_CEILAP-BA.terra.1km (accessed in July 2016).

sampling stations according to the hydrological behavior and seasonal variation. Coastal zones influenced by river discharges show gradients in nutrient concentrations related with changes in salinity [42]. Both TN and TP concentrations were positively associated with the RdlP and Parana flows and negatively associated with salinity and Secchi depth. In addition, the stations with the highest nutrient concentrations were associated to the axis 1 of the PCA, which represents the hydrological variability of the system. These results suggest that an increase in the RdlP flow promotes an increase of nutrient availability in the MCZ, particularly during the EN phase.

On the other hand, high Chl *a* concentrations and temperature associated with PCA axis 2 are interpreted as a seasonal variation of biomass phytoplankton related to temperature. In [28], phytoplankton Chl *a* usually displays a concentration peak by the end of summer and the beginning of autumn, showing a unimodal trend in the seasonal pattern. There are few studies of the RdlP regarding the seasonal patterns in total chlorophyll concentration; nevertheless, recent studies have demonstrated that salinity, depth, and light availability, which attain maximum values in spring, are the main variables controlling the biomass and composition of the phytoplankton community in the RdlP [43].

3.4. Weather, hydrological, and oceanographic interactions

Significant correlations between Niño indexes (Niño 3.4 > Niño 4 > Niño 1 + 2 > Niño 3) and flows of RdlP, Uruguay, and Parana rivers were only found during EN event (n = 10; **Table 1**).

During the "EN" phase, we observed significant correlations between the RdlP and Uruguay River flow and the different Niño indexes (Niño 1 + 2, Niño 3, Niño 3.4, and Niño 4), where Niño 3.4 showed the highest correlation coefficients. Similar correlation coefficients (r = 0.51) were observed between Niño 3.4 and the Uruguay River flow [19]. The associations found in this study with the RdlP and Uruguay River flow were higher than those of the above

	Q RdlP	Q Uruguay river	Q Paraná river
Niño 1 + 2	−0.685 (P = 0.029*)	NS	−0.842 (P = 0.002**)
Niño 3	NS	0.673 (P = 0.033*)	NS
Niño 3.4	0.888 (P = 0.001**)	0.912 (P = 0.000**)	0.578 (P = 0.080 MS)
Niño 4	0.806 (P = 0.005**)	0.782 (P = 0.008**)	0.648 (P = 0.043*)

P = * 0.05; ** <0.01; MS: marginally significant; NS: not significant.

Table 1. Correlations (Rs) between Niño indicators and flow (Q) of RdlP, Uruguay and Parana rivers.

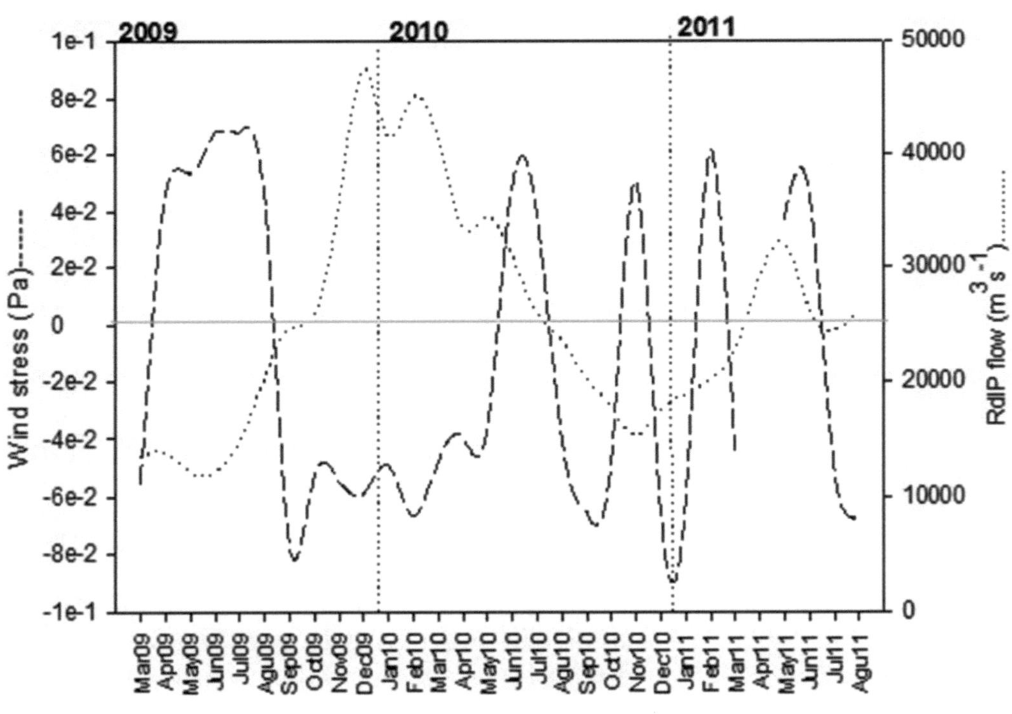

Figure 10. Wind stress fluctuations (Pa) (dashed lines) and RdlP flow ($m^3 s^{-1}$).

mentioned authors. During ENSO events, joint monitoring of the Niño 3.4 index and the RdlP and the Uruguay River flow would allow the assessment of the temporal variability in hydrology of the system, promoting the generation of early warnings of floods or maximum water levels in both systems.

The wind stress and flow of the RdlP showed in phase and out of phase fluctuations (*sensu* [23]) (**Figure 10**). Wind stress and RdlP flow (R_s $(_{0.021,\ 29})$ = −0.41) and Uruguay River flow (R_s $(_{0.001,\ 29})$ = −0.50) showed significant inverse correlations. Minimum continuous stress values were observed during September 2009 and April 2010, coinciding with peak flows of RdlP, while maximum values of stress (May and August 2009) coincided with the RdlP minimum flows.

Flow (RdlP, Uruguay and Parana rivers) and average salinity showed inverse linear relationships with higher coefficients recorded during the EN phase than during "LNN" phase (**Figure 11**).

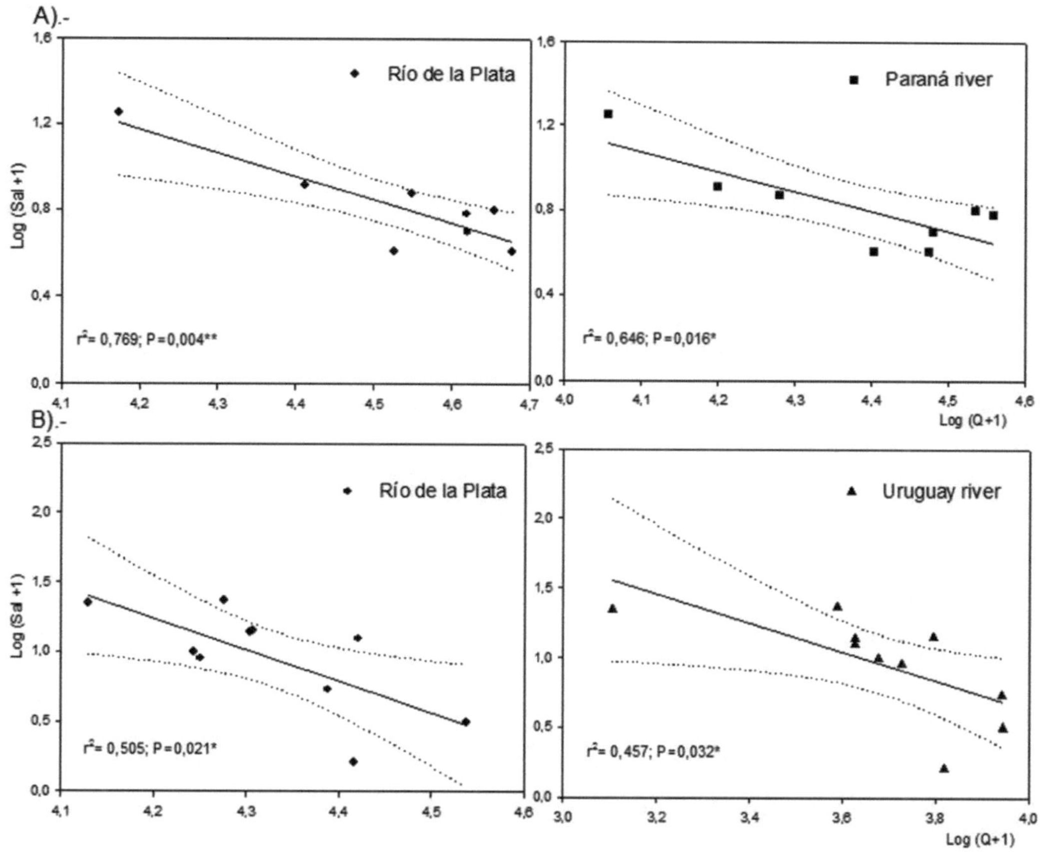

Figure 11. Inverse linear relationships between MCZ average salinity and RdlP, Uruguay and Parana rivers flow. (A) "El Niño," n = 8; (B) "La Niña-Neutral," n = 10.

Wind stress and discharge levels are the main processes responsible for extending the estuarine discharge plume of the RdlP over the Atlantic continental shelf [23]. In this system, the E-ESE-E wind direction promotes the inflow of ocean water into the estuary, while W-WNW winds promote discharge of affluent rivers (Parana and Uruguay rivers), increasing or reducing average salinity and the upriver or downriver displacement of the saline front [18, 19]. During high discharge periods (Q RdlP < 30,000 $m^3 s^{-1}$), which are largely associated to ENSO events (5 in 7 between 1959 and 1990), only the event in 1992 presented wind stress conditions favorable for the dispersion of the river plume into the internal zone of the RdlP. In the remaining events, the wind strength opposite to the penetration of the plume was negligible [23]. During September 2009–May 2010 (EN phase), we observed "out of phase" fluctuations between the RdlP flow and wind stress, while during "LNN" fluctuations were mostly of the "in phase" type (*sensu latu* [23]). According to [23], high flows in the Uruguay and Parana rivers combined with minimum wind stress promote optimal conditions for the discharge and penetration of RdlP waters into the Uruguayan and Argentinean coastal zone. Similar results were observed for the RdlP discharge plume over the continental shelf [44]. In this study, we recorded a predominance of flow over wind stress as the main driver for the extension of the plume discharge over the north coast of the RdlP estuary during EN phase.

4. Conclusions

During the 2009–2010 ENSO event, in the EN phase, there is an increase in rainfall from spring 2009, promoting an increase in the flows of the main tributary rivers to the RdlP estuary. During EN phase, the RdlP flow and wind stress are the principal drivers of oceanographic condition at the north coast of RdlP.

Acknowledgements

We thank PEDECIBA, ANII (SNI), UdelaR (CSIC-DT), and Intendencia de Montevideo for their financial support. We also thank the SOHMA (Service Oceanography, Hydrography and Meteorology of the Navy, Uruguay) staff by the wind records, and Mario Bidegain by the flow data. B.S. in Biol. Pablo Mones for the English translation.

Author details

Ernesto Brugnoli[1]*, José Verocai[1], Pablo Muniz[1] and Felipe García-Rodríguez[2]

*Address all correspondence to: ebo@fcien.edu.uy

1 Oceanography and Marine Ecology Laboratory, Science Faculty, University of the Republic, Montevideo, Uruguay

2 CURE, University of the Republic, Rocha, Uruguay

References

[1] McPhaden MJ. Genesis and evolution of the 1997-1998 El Niño. Science. 1999;**283**:950-954

[2] Lee T, McPhaden MJ. Increasing intensity of El Niño in the central equatorial Pacific. Geophysical Research Letters. 2010;**37**:L 14603. DOI: 10.1029/2010GL044007

[3] Barros VR, Grimm AM, Doyle ME. Relationship between temperature and circulation in south-eastern South America and its influence from El Niño and La Niña events. Journal Meteorology of Society Japan. 2002;**80**:23-32

[4] Feidler P. Environmental change in the eastern tropical Pacific Ocean review of ENSO and decadal variability. Marine Ecology Progress Series. 2002;**2443**:265-283

[5] Lavaniegos BE, Jiménez-Pérez LC, Gaxiola-Castro G. Plankton response to El Niño 1997-1998 and La Niña 1999 in the southern region of the California current. Progress in Oceanography. 2002;**54**:33-58

[6] Sellanes J, Quiroga E, Neira R, Gutiérrez CD. Changes of macrobenthos under different ENSO cycle conditions on the continental shelf off central Chile. Continental Shelf Research. 2007;**27**:1002-1016

[7] Defeo O, Castrejón M, Ortega L, Kuhn AM, Gutiérrez NL, Castilla JC. Impacts of climate variability on Latin American small-scale fisheries. Ecology and Society. 2013;**18**(4):30-35

[8] Caffera M, Berbery EH. Climatología de la Cuenca del Plata. In: Barros V et al., editors. El Cambio Climático en la Cuenca del Plata. 1st ed. Buenos Aires, Argentina: CONICET; 2006. pp. 19-47

[9] Grimm A, Barros VR, Doyle ME. Climate variability in southern South America associated with El Niño and La Niña events. Journal of Climate. 2000;**13**:35-58

[10] Barreiro M. Influence of ENSO and the South Atlantic Ocean on climate predictability over South-eastern South America. Climate Dynamics. 2010;**35**:1493-1508

[11] Robertson AW, Mechoso CR. Interannual and decadal cycles in river flows of South-eastern South America. Journal of Climate. 1998;**11**:2570-2581

[12] Berbery EH, Doyle M, Barros V. Tendencias regionales en la precipitación. In: Barros V et al., editors. El Cambio climático en la Cuenca del Plata. 1st ed. Buenos Aires, Argentina: CONICET; 2006. pp. 67-78

[13] Wolff M, Ruiz J, Taylor M. El Niño induced changes to the Bolivar Channel ecosystem (Galapagos): Comparing model simulations with historical biomass time series. Marine Ecology Progress Series. 2012;**446**:7-22

[14] Bollens SM, Breckenridge JK, Vandenhooff RC, Cordell JR. Mesozooplankton of the lower San Francisco estuary: Spatio-temporal patterns, ENSO effects and the prevalence of non-indigenous species. Journal of Plankton Research. 2011;**33**:1358-1377

[15] Camillioni I, Barros V. The Paraná River response to El Niño 1982-83 and 1997-98 events. Journal of Hydrobiology. 2000;**1**:412-430

[16] García-Rodríguez F, Brugnoli E, Muniz P, Venturini N, Burone L, Hutton M, Rodríguez M, Pita A, Kandratavicius N, Pérez L, Verocai J. Warm phase ENSO events modulte the continental freshwater input and the trophic state of estuarine sediments: An example from a large South America estuary. Marine and Freshwater Research. 2014;**65**(1):1-11

[17] Pérez L, García-Rodríguez F, Hanebuth TJ. Variability in terrigenous sediment supply offshore of the Río de la Plata (Uruguay) recording the continental climatic history over the past 1200 years. Climate of the Past. 2016;**12**:623-634

[18] Nagy G, Pshennikoy-Severova V, Robatto P. Monthly variation of salinity in the frontal zone of the Río de la Plata off Montevideo, in response to consecutive ENSO fluctuations and the flow of the Uruguay River (1998-2000). In: Vizziano et al., editors. The Río de la Plata Research to Manage the Environment, Fish Resources and the Fishery in the Saline Front. Montevideo, Uruguay: EcoPlata Program; 2002. pp. 21-30

[19] Nagy G, Severov D, Pshennikov-Severova V, De los Santos M, Lagomarsino J, Sans K, Morozov E. Río de la Plata estuarine system: Relationship between river flow and frontal variability. Advances in Space Research. 2008;**41**:1876-1881

[20] Nagy G, Seijo L, Verocai J, Brugnoli E, Bidegain M. Enfoque, conocimiento y medidas para enfrentar las amenazas del clima presente en la zona frontal del Río de la Plata, Uruguay. Revista Iberoamericana Manejo Costero Integrado. 2013;**1**(2):69-87

[21] Severov D, Mordecki E, Pshennikov-Severova V. SST anomaly variability in South-western Atlantic and El Niño/Southern Oscillation. Advances in Space Research. 2004;**33**:343-347

[22] Guerrero RA, Acha E, Framiñan MB, Lasta CA. Physical oceanography of the Río de la Plata estuary, Argentina. Continental Shelf Research. 1997;**17**(7):727-742

[23] Piola A, Matano RP, Palma ED, Möller O, Campos EJ. The influence of the Plata River discharge on the South Atlantic shelf. Geophysical Research Letters. 2005;**32**. DOI: LO1603, 10.1029/2004GL021638

[24] Acha EM, Mianzán H, Guerrero R, Carreto J, Giberto D, Montoya N, Carignan M. An overview of physical and ecological processes in the Río de la Plata estuary. Continental Shelf Research. 2009;**28**:1779-1588

[25] Framinñan M, Brown OB. Study of the Río de la Plata turbidity front, part I: Spatial and temporal distribution. Continental Shelf Research. 1996;**16**(10):1259-1282

[26] FREPLATA. Análisis Diagnóstico Transfronterizo del Río de la Plata y su Frente Marítimo. 1st ed. Montevideo, Uruguay: FREPLATA; 2005

[27] Brugnoli E, Muniz P, Venturini N, Burone L. Environmental perturbation and coastal benthic biodiversity in Uruguay. In: Willis IC, editor. Progress in Environmental Research. 1st ed. New York: Nova Publishers; 2007. pp. 75-126

[28] Gómez-Erache M, Vizziano D, Muniz P, Nagy G. The health of the Río de la Plata system: Northern coast Uruguay. In: Chopin, T and Wells PG, editors. Opportunity and Challenges for Protecting, Restoring an Enhancing Coastal Habitats in the Bay of FundyProceed. Of the Bay of Fundy Science; Saint John, New Brunswick. 2001. p. 17-35

[29] Muniz P, Danulat E, Yanicelli B, García-Alonso J, Medina G, Bícego MC. Assesment of contamination by heavy metals and petroleum hydrocarbons in sediments of Montevideo harbour (Uruguay). Environmental International. 2004;**29**:1019-1028

[30] Muniz P, Venturini N, Martins C, Bano-Munsh A, García-Rodríguez F, Brugnoli E, Lindroth-Dauner AL, Caruso-Bícego M, García-Alonso J. Integrated assessment of contaminants and monitoring of an urbanized temperate harbour (Montevideo, Uruguay): A 12-year comparasion. Brazilian Journal of Oceanography. 2015;**63**(3):311-330

[31] Venturini N, Bícego MC, Taniguchi S, Sasaki ST, García-Rodríguez F, Brugnoli E, Muniz P. Multi-molecular marker assessment of organic pollution in shore sediments from the Río de la Plata estuary. Marine Pollution Bulletin. 2015;**91**:461-475

[32] American Public Health (APHA). Standard Methods for the Examination of Water and Waste Water. 21st ed. New York ed. 2005

[33] Monteiro MI, Ferreira FN, de Oliveira NM, Avila AK. Simplified version of the sodium salicylate method for analysis of nitrate in drinking water. Anayltica Chemica Acta. 2003; **477**:125-129

[34] Valderrama JC. The simultaneous analysis of total nitrogen and total phosphorus in natural waters. Marine Chemistry. 1981;**10**:109-122

[35] ter Braak CJF, Smilauer P. CANOCO Reference Manual and CanoDraw for Windows user's Guide: Software for Canonical Community Ordination. 1st ed. Ithaca, New York: Microcomputer Power; 2002

[36] Kim WM, Yeh SW, Kim JH, Kug JS, Kwon MH. The unique2009–2010 El Niño event: A fast phase transition of warm pool El Niño to La Niña. Geophysical Research Letters. 2011;**38**. DOI: L15809, 10.1029/2011GL048521

[37] National Oceanographic and Atmospheric Administration (NOAA). ENSO: Recent evolution, current status and predictions. Update prepared by Climate Predictor Center/ NCEP [Internet]. 2011. Available from: http://www.cpc.ncep.noaa.gov/products/analysis_monitoring/lanina/enso_evolution-status-fcsts-web.pdf [Accessed: 03/2016]

[38] Ashok H, Beheran SK, Rao SA, Weng H, Yamagata T. El Niño Modoki and its possible teleconnection. Journal of Geophysical Research. 2007;**112**(C11007). DOI: 10.1029/2006JC003798

[39] Comisión Administradora del Río de la Plata (CARP). Estudio para la Evaluación de la Contaminación del Río de la Plata. Montevideo, Uruguay; 1990

[40] Nagy G, Martínez C, Caffera CM, Pedrosa G, Forbes EA, Perdomo A, López-Laborde J. The hydrological and climatic setting of the Río de la Plata. In: Wells PG, Daborn GR, editors. The Río de la Plata. An Environmental Overview. An EcoPlata Project Background Report. 1st ed. Halifax, Nova Scotia, Canada: Dalhousie University; 1997. pp. 17-68

[41] Kruk C, Martínez A, Nogueira L, Alonso C, Calliari D. Morphological traits variability reflects light limitation of phytoplankton production in a highly productive subtropical estuary (Río de la Plata, South America). Marine Biology. 2014;**162**(2):331-341

[42] Marchovecchio J, Freije HR. Procesos químicos en estuarios. 1st ed. Bahía Blanca, Argentina: Universidad Tecnológica; 2013

[43] Martínez A, Ortega L. Delimitation of domains in the external Río de la Plata estuary, involving phytoplanktonic and hydrographic variables. Brazilian Journal of Oceanography. 2015;**63**(3):217-228

[44] Machado I, Barreiro M, Calliari D. Variability of chlorophyll-a in the southwestern Atlantic satelite images: Seasonal cycle and ENSO influences. Continental Shelf Research. 2013;**53**: 102-109

Chapter 3

Pesticides in Worldwide Aquatic Systems: Part I

Catarina Cruzeiro, Eduardo Rocha and
Maria João Rocha

Additional information is available at the end of the chapter

http://dx.doi.org/10.5772/intechopen.71644

Abstract

The occurrence of pesticides in aquatic environments is registered worldwide, but few or no approaches have been used to summarize and integrate the data. In this work, 30 countries and 95 aquatic systems were taken into consideration, using the data collected in the past 17 years. Data were evaluated by continent, with a special focus on Europe, as the continent with the most information available. However, in terms of analyzed pesticides, the insecticides were the most common category of pesticides being applied in excess in several Asian countries. Moreover, priority pesticides settled for elimination were/are still present in almost all the continents, demonstrating that those compounds continue to be used. This leads to the existence of environmental mixtures containing both legal and illegal pesticides, which are able to affect different trophic levels, including humans. Thus, action plans like international discussions and pacts should exist to regulate the adequate usage of pesticides, and a continuous environmental monitoring should be enforced to understand potential toxicological risks promoted by these compounds. Further considerations, based on the Stockholm Convention list and European Directive 2013/39/EU as references, were used to evaluate the degree of contamination in the studied aquatic systems.

Keywords: insecticides, herbicides, fungicides, water, estuaries, 2013/39/EU, Stockholm Convention

1. Preamble

The current overuse and abusive application of pesticides may impact diverse aquatic ecosystems in both the short and long term. Due to their physicochemical properties, pesticides can circulate through various mechanisms, converting into an additional source of contamination to aquatic environments, mainly the estuaries. Although many scientific and governmental works have been published to alert to these facts, poor approaches have been used to connect all

INTECH
open science | open minds

© 2018 The Author(s). Licensee InTech. This chapter is distributed under the terms of the Creative Commons Attribution License (http://creativecommons.org/licenses/by/3.0), which permits unrestricted use, distribution, and reproduction in any medium, provided the original work is properly cited.

available data. With this in mind, the main goal of this work is to review a significant amount of published representative data from a variety of aquatic systems, including rivers, estuaries, and coastal areas, and discuss the published results, around the world, taking into consideration factors such as geographic variability (continental and regional), matrices, pesticide category, and the European legislation.

Due to the volume of available information, the review is restricted to a period of 17 years (from 2000 to 2017) of publications. All the available data—average minimum (av-min), average maximum (av-max), and average of averages (av-av) concentrations—were collected and expressed as ng/L. Data were grouped by pesticide category. Europe is used as the main pillar of this study because it is the continent with the largest amount of data available. Online databases, as Web of Science (Thomson Reuters) and PubMed (NCBI), were used to access the indexed articles used in this work.

2. Water matrix

Eighty-eight articles were reviewed and compiled in **Table 1**. Matrices such as surface waters and dissolved aqueous phase represent a total of 79 and 6% of the collected data, respectively. Among these, 62% of the analyzed data refers to Europe, and the rest is divided between Africa and Asia (each with 13 and 18%, respectively), followed by South America and Oceania. No data were found for North America and Antarctica with the above presented criteria (**Table 1**); thus, when citing herein "worldwide", these continents are not included. Fifty aquatic systems were studied in Europe, from which Spain stands out with 13 (published in 19 journals).

Overall, the data collected between 1993 and 2017 show average concentrations of pesticides ranging from ~17 to ~9936 ng/L (**Table 1**). Among the selected articles, 141 compounds were detected and quantified in Europe, 57 in Asia, 42 in Africa, 21 in Oceania, and 33 in South America. The highest average concentrations and standard deviations (SD) were measured in Asia (875 ng/L; SD 3468), followed by Europe (638 ng/L; SD 10761), South America (487 ng/L; SD 2448), Africa, and finally, Oceania (230 ng/L; SD 1500).

On a worldwide scale, the insecticides prevail (60%) in terms of available and quantified data when compared with both herbicides (33%) and fungicides (7%). Per continent, the percentage of insecticides are more than 90% in Africa and Asia, summing approximately 45% in Europe, 71% in South America, and 19% in Oceania. No cases were reported in North America and Antarctica (**Figure 1**). While the high percentage of insecticides in Asia may be due to the high cereals production (more than 13×10^8 tonnes), in Africa, it can be linked to cereals production, plague control, and vector-borne diseases control [86–88]. In South America, studies alert to abusive usage of insecticides for pest control due to resistant species and the introduction of nonnative ones [89, 90]. In **Figure 1**, a peculiar different pattern is observed for the percentages of types of pesticides in Europe versus Oceania, which for the first case may be due to the high number of compounds quantified (141) or, more likely, to a response to diverse agriculture practices and industrial needs [91].

Continent/country	Number of aquatic systems	Quantified pesticides	Sampling year	av-min	av-max	av-av	Reference
				ng/L			
Africa							
Benin	1	6	2010	138.7	358.0	224.9	[1]
Egypt	2	12–13	1993	0.1	0.2	0.1	[2]
Ghana	2	4–11	2004	0.3	0.9–120.5	0.1–97.3	[3, 4]
Kenya	1	2	na	na	na	9375	[5]
Mozambique	1	16	na	24	43.4	30.6	[6]
Nigeria	5	1–14	2014	405.5–1930	431.0–3267	190.0–2163	[7–9]
South Africa	5	4–15	1999–2002	na–25	na–135	35.2–77.9	[10, 11]
Asia							
China	11	5–30	1999–2014	0.3–794.3	4.6–31,261	1.5–7384	[12–22]
India	3	3–13	2009–2015	0–2133	2.2–194,700	0.2–13,166	[23–25]
Macau	1	18	2001	0.8	3.5	1.6	[26]
Russia	2	7	2003–2005	na	na	0.1	[27]
Vietnam	1	13	2012	na	2246	398.5	[28]
Europe							
Central and Eastern Europe	1	9	2007	na	24.1	6.3	[29]
Belgium	2	6–7	2002–2004	na	na	48.4–312.1	[30–31]
Bulgaria	1	8	na	6.6	10.4	5.3	[32]
France	6	3–19	2003–2010	81.7–317.4	94.3–3452	26.9–566.7	[33–37]
Germany	6	1–19	2001–2003	4	250–5600	9.1–580	[38–39]
Greece	7	3–23	1996–2007	11.6–47.3	29.2–803.3	19.6–99.3	[40–43]
Hungary	1	2	2010	na	na	417.1	[44]
Italy	1	9	2008	1.2	4.4	1.9	[45]
Norway	1	12	2014	0.1	0.6	0.3	[46]
Poland	2	8–12	2002–2003	1.3–525.4	55.6–1323	8.5–42	[47–48]
Portugal	7	8–48	2004–2012	5.9–6487	125–290,345	31.2–17,667	[49–57]
Romania	3	7	2004–2013	8.3	9.8–39.7	1.6–37.1	[58–59]
Spain	13	1–45	1996–2013	6.1–58.4	35.8–947	4–940	[60–74]
The Netherlands	na	13	2008	34.6	79.2	43.8	[75]
Oceania							
Australia	5	4–10	2006–2010	1.5–138.3	8.5–3399	2.8–759.1	[76–80]

Continent/country	Number of aquatic systems	Quantified pesticides	Sampling year	av-min	av-max	av-av	Reference
				ng/L			
South America							
Argentina	2	3–8	2012	20–28.3	139.6–3783	53.5–323.3	[81–82]
Brazil	2	10–11	1999–2005	4.9–18.1	40.1–50.6	12.9–23.6	[83–84]
Chile	1	8	2013–2014	na	na	2.6	[85]

Table 1. Pesticide concentrations [average minimum (av-min), average maximum (av-max) and average of averages (av-av) values; ng/L] in water samples, displayed by continent, country, and aquatic system; the number of quantified pesticides and sampling year were also added (na: not applicable).

Looking at the nature of the matrices, while most studies have been using surface water as target (78%), the rest have been tackling groundwater (9%), dissolved aqueous phase (6%), and even others (**Figure 1**).

In spite of these facts, we should be aware that these results are dependent on the authors' selection, which may not correspond entirely to what is present in the aquatic systems.

Amid continents, the number of quantified compounds was similar (~12) with the exception of Europe, which presented a higher number of measured pesticides (~23) and a higher number of aquatic systems monitored.

The quantified pesticides data were also compared to the average and maximum levels set by Directive 2013/39/EU. Considering this aspect, the pesticides with levels above those established by the Directive are referred herein as positive cases (**Table 2**). A higher number of positive cases were registered for average concentrations (with percentages ranging from 31 to 75%) than for maximum concentrations (with percentages ranging from 12 to 39%). Considering both average and maximal concentrations, higher percentages of pesticides considered dangerous and banned by the Directive 2013/39/EU were registered in Asia (mainly China) and in South America. However, in South America (mainly Brazil), several pesticides that are legally forbidden in Europe (at least in European Union) still are legal in South America. The last observation leads to an over usage of these compounds in the respective region. In Asia (China), dicofol (structural similarity to DDT) will become forbidden in 2018 by the governmental agencies.

Insecticides are the only common pesticide category among continents, demonstrating its value in agriculture and urban gardening. The previous scenario, ruled by Asia and South America, is now changed, where Europe presents almost the double (n = 74) of positive cases (for average concentrations), when compared to Africa (n = 39) and Asia (n = 41). Few cases were observed in other continents. This denotes the importance of the European legislation and how far we are to accomplish its goals.

In Europe, pesticide levels averaged between ~4 and ~399 ng/L. Herein, the top three countries with published articles (from a total of 42 publications) are Spain (30%), Portugal (26%), and Greece (9%). These three countries reported the presence of more than 79 (Spain), 94 (Portugal),

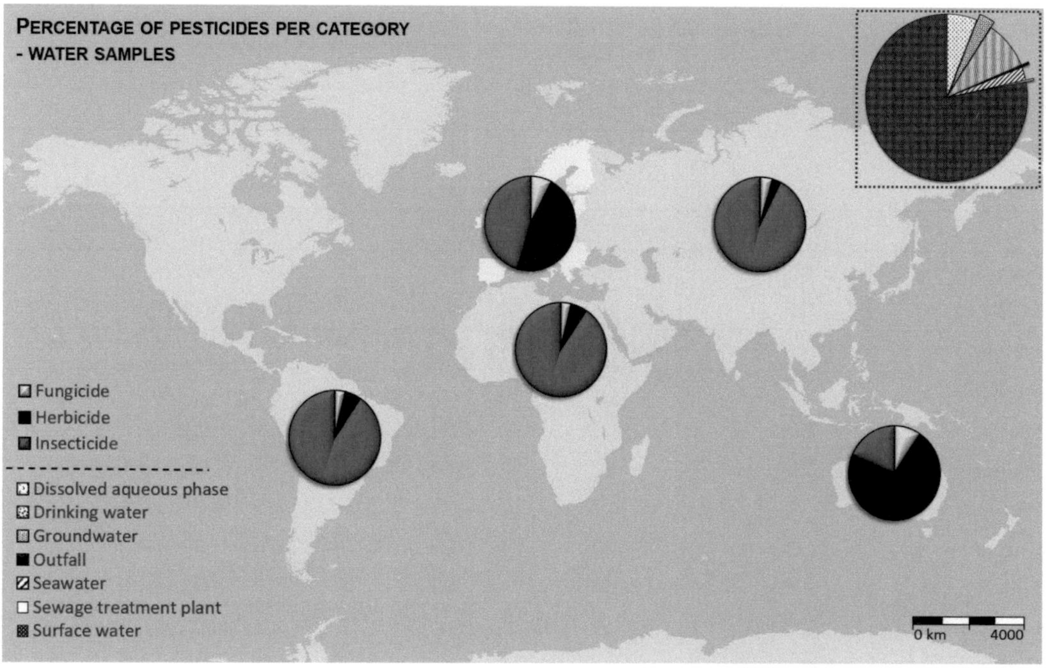

Figure 1. Representation of the quantified pesticides in water samples (%), per category, on each continent; the right upper corner figure represents the type of matrices found worldwide.

and 26 (Greece) pesticides in different aquatic systems. Looking at the number of positive cases, for average and maximum Directive established limits, Portugal (n = 74) stands out when compared to Spain (n = 10) and Greece (n = 28) [49, 51–57, 92]. These results demonstrate that the Portuguese aquatic systems are loaded with extreme high concentrations of pesticides, which can be due ineffective water treatment and/or abusive usage of pesticides along the water courses. It should be noted that the main rivers such as Minho, Douro, and Tagus have their origin in Spain, which can also contribute to the high levels observed in Portugal.

Due to the different number of compounds analyzed per published articles, the most frequent pesticides (more than 10 observations, i.e., quantification of pesticides in different aquatic systems or countries) were re-analyzed to compare the average concentrations between the different continents. The majority of the quantified pesticides (**Table 3**) belong to the priority list of persistent organic pollutants [94, 95]. Among these substances, which were settled in the Stockholm Convention list to be eliminated, the hexachlorocyclobenzene (HCB), DDT, aldrin, dieldrin, endrin, and hexachlorocyclohexane (HCH) were quantified in almost all the continents even after 2001, showing a continuous usage of these illegal substances. The same was registered for DDT, heptachlor, and hexachlorocyclobenzene after 2009. In fact, while HCB, aldrin, and dieldrin were measured in higher average concentrations in Europe, DDT and HCH were more prominent in Asia and endrin, endosulfan, and heptachlor were quantified in higher amount in Africa. In South America, the levels of the banned compounds were not particularly high; nonetheless, further studies should be undertaken to confirm the published data.

Continent/country	Average amounts (ng/L)	Number of cases	Samples above 2013/39/EU		References
			av	max	
Africa					
Fungicide	32.5	6	4	0	[2, 10]
Insecticide	312.9	51	39	7	[1–4, 6–8, 10, 11]
Asia					
Fungicide	4.1	3	1	0	[12, 27]
Insecticide	2270	72	41	29	[12–27]
Europe					
Fungicide					
Greece	72.5	2	2	1	[41]
Portugal	45.0	8	6	1	[50, 53, 55–57, 92]
Herbicide					
Belgium	243.3	8	1	0	[30, 31]
France	294.6	10	4	3	[34, 36, 37]
Germany	73.5	19	2	3	[38, 39]
Greece	49.8	8	1	1	[40, 42, 43]
Portugal	370.6	32	7	4	[49–54, 56, 57, 92]
Spain	125.0	59	2	2	[61, 63, 65–72, 74]
Insecticide					
Belgium	56.0	1	1	0	[30]
Bulgaria	17.2	3	2	1	[32]
France	140.1	5	5	2	[34–36]
Greece	48.5	16	14	9	[40, 41]
Italy	2.8	2	1	1	[45]
Poland	47.2	6	4	2	[47, 48, 93]
Portugal	398.5	54	42	14	[49–51, 53–57]
Romania	4.1	5	2	2	[58, 59]
Spain	9.0	20	3	3	[62–64, 66–69, 74]
Oceania					
Herbicide	503.1	11	3	3	[76–80]
Insecticide	2.0	2	1	1	[76, 77]
South America					
Insecticide	112.1	11	8	6	[82–85]

Table 2. Pesticides average (av) and maximum (max) concentrations (ng/L) in water samples, displayed by continent and pesticide category; the number of quantified pesticides, as well as the number of samples above 2013/39/EU Directive levels, were also included; Europe is presented with more detailed information; references are only defined for the samples above the 2013/39/EU Directive, per category.

Higher average concentrations of the same order of magnitude in Africa and Europe (global average ~38 ng/L) and lower amounts in Asia (~4 ng/L) were registered for the fungicide HCB. Herbicides such as atrazine and simazine were measured in Europe, Oceania, and South America, where the highest average concentrations were observed for the first two continents. Among herbicides, diuron stands out with concentrations 6-fold higher in Oceania (~1200 ng/L), when compared to the other continents (~200 ng/L). Among insecticides, ∑DDT, ∑cyclodiene, chlorpyriphos, ∑endosulfan, ∑heptachlor + heptachlor epoxide, ∑HCH, and malathion were most frequent in Africa, Asia, Europe, and South America. Comparing the total average sum of these insecticides (∑), Asia had the highest concentrations (~10,000 ng/L), followed by Africa (~3000 ng/L), Europe (~1800 ng/L), and finally, South America (~300 ng/L). The extremely high values in Asia are due to punctual observations in the Deomoni River (India) and in the Yellow River (China), which do not reflect the average concentration in Asia [21, 24]. However, when considering all pesticides from **Table 3**, we recorded similar concentrations (from ∑ ~7460 ng/L to ∑ ~10,540 ng/L) among Africa, Asia, and Europe, confirming that high punctual concentrations occur in different continents. The high concentrations reported for chlorpyriphos (in Asia and South America) and for diazinon (in Africa) are above the LC_{50} and/or EC_{50} observed in short-time exposures (48–96 h), for fish (as the rainbow trout) and invertebrates (as the crustaceans daphnia and mysid shrimps). Individually, these compounds can already cause mortality to 50% of the exposed population; however, a worst-case scenario may occur if these compounds are present in an environmental mixture (further considerations are done in chapter *Pesticides in Worldwide Aquatic Systems: Part II*).

The parent compound/residues ratios were calculated for DDT, endosulfan, and heptachlor. Results demonstrate an active use of DDT in Asia (1.4), while for endosulfan and heptachlor, the active use is spread among diverse continents (Africa, Asia, Europe, and South America).

The most frequent pesticides (equal or more than 10 quantifications in different aquatic systems or countries) were selected and grouped by category for the European countries (**Table 4**), reaching 23 compounds. The concentrations of eleven of these pesticides are above the Maximum Residue Limits (MRLs) set by 2013/39/EU Directive. The range of concentrations (min-max) was assessed, displaying the most substantial differences between countries. Seven pesticides (alachlor, aldrin, dieldrin, chlortoluron, dimethoate, diuron, and terbuthylazine) stand out with the highest ranges (numbers in bold, **Table 4**). Alachlor is present in the Iberian Peninsula at levels above the 2013/39/EU Directive limits set for average concentrations in surface waters, which may relate to a regional application of this herbicide [49–51, 53–57, 60, 63, 66, 67, 71, 72]; the same was observed for diuron, in Spain, France, and Belgium [30, 34, 37, 60, 63, 67, 70, 72]. The cyclodiene pesticides (∑aldrin and dieldrin) were above the annual average concentrations (∑ ~5 ng/L) set by the same directive for all registered cases, presenting extremely high amounts in Portugal (∑cyclodienes ~2174 ng/L), demonstrating an abusive and illegal use of these compounds in these regions [50, 51, 53, 55, 57]. Remarkably, none of these pesticides were above the LC_{50} and/or EC_{50} documented for the most typical organisms representative of the various trophic models.

Average amounts (ng/L)	Africa	Asia	Europe	Oceania	South America	References
Fungicide						
HCB	32.5	4.1	43.0			[2, 10, 12, 27, 32, 41, 50, 53, 55, 56]
Herbicide						
Alachlor		1.7	529.9		11.0	[34, 36, 38, 40, 49–57, 60, 63, 66, 67, 71, 72, 83]
Atrazine	107.9		138.9	482.9	17.0	[34, 36, 38, 40, 49–57, 60, 63, 66, 67, 71, 72, 83]
Atrazine-desethyl	6.0		173.2	65.2		[6, 29, 33, 38, 40, 43, 50, 53–57, 61, 63, 66, 67, 69, 71, 72, 78, 80, 96]
Chlortoluron			68.0			[34, 36, 38, 49, 63, 67, 72, 75]
Diuron	200.0		239.7	1200		[6, 29, 30, 37, 38, 46, 49, 63, 67, 70–72, 74, 78–80, 97]
Isoproturon			34.7			[29, 30, 36, 38, 46, 63, 67, 72, 74, 75, 97]
Metolachlor		22.7	57.4		5.0	[12, 30, 33, 34, 36, 38, 46, 49, 50, 52–54, 56, 63, 65–67, 71, 72, 74, 83]
Simazine	9.0		88.8	120.3	9.0	[6, 29–31, 38, 40, 43, 49, 50, 53, 55, 56, 61, 63, 65–68, 70–72, 74, 76, 77, 79, 80, 83]
Terbuthylazine	27.5		280.2			[6, 29, 31, 38, 46, 49, 50, 53–57, 61–63, 65–67, 69–72]
Terbutryn			98.9	5.0		[37, 39, 53, 55–57, 66, 69, 76]
Trifluralin		4.5	133.6		7.0	[12, 34, 36, 40, 42, 53, 55, 56, 71, 83]
Insecticide						
ΣDDT	744.7	1765	122.9		105.0	
2,4′-DDD	138.8	25.1	31.0			[1, 7, 10, 23, 25, 26, 64]
2,4′-DDT	212.7	369.0	4.0		6.0	[7, 10, 12, 15, 17, 20–23, 25, 26, 64, 84]
4,4′-DDD	103.3	132.7	15.2		41.0	[1, 2, 4, 7, 10, 12, 14, 15, 17, 19–21, 23, 26, 35, 50, 53, 55, 56, 59, 62, 64, 68, 84, 93]

Average amounts (ng/L)	Africa	Asia	Europe	Oceania	South America	References
4,4′-DDE	139.9	679.0	18.8		36.0	[1–4, 7, 12–14, 16, 17, 19, 20, 22, 23, 26, 41, 54–56, 64, 68, 84, 93]
4,4′-DDT	150.0	559.2	54.0		22.0	[1–3, 7, 10, 12–17, 19–23, 25, 26, 32, 35, 41, 47, 50, 53, 55, 56, 59, 64, 68, 84, 93]
DDT/DDE + DDD	0.9	1.1	0.9		0.4	
ΣCyclodiene	334.8	48.4	1324		1.5	
Aldrin	251.3	16.6	392.1		1.5	[4, 7, 8, 10, 12–14, 16, 17, 19, 20, 25, 26, 32, 41, 48, 50, 51, 53, 55, 56, 68, 85]
Dieldrin	44.5	12.7	898.7	3.0		[4, 6, 8, 10, 12, 19, 20, 25–27, 30, 41, 48, 50, 51, 57, 68, 76, 81]
Endrin	39.0	19.1	32.8			[8, 10, 12, 13, 16, 17, 19, 20, 25, 26, 32, 48, 50, 53, 55, 56, 68]
Chlordane γ	24.9	4.0	3.9			[2, 4, 10, 12, 13, 53, 57, 59, 72, 73, 83]
Chlorpyrifos	2.6	3103	14.9		110.0	[6, 12, 25, 40, 45, 50, 53, 55–57, 66, 69, 71, 74, 82]
Diazinon	4040		39.4			[5, 6, 37, 40, 42, 43, 45, 47, 49, 50, 53, 55–58, 63, 67, 69, 72, 74]
Dimethoate		360.0	4304	2.0	35.0	[18, 40, 45, 46, 49, 50, 53–57, 67, 69, 72, 74, 76, 81]
ΣEndosulfan	103.3	50.3	112.2		33.3	
Endosulfan α	77.8	15.0	87.1		10.8	[6, 8, 11, 12, 16, 19, 25–27, 32, 41, 50, 53, 55–57, 59, 64, 83–85]
Endosulfan β	25.5	35.4	25.0		22.5	[4, 6, 8, 12, 13, 16, 19, 26, 41, 50, 53, 55–57, 64, 83, 84]
Endosulfan sulfate	22.6	41.4	40.8		7.0	[4, 6, 8, 11, 13, 16, 19, 25, 41, 53–57, 83, 84]
Endosulfan/Endosulfan sulfate	4.6	1.2	2.8		4.8	
Fenitrothion			77.5			[24, 40, 45, 47, 50, 53, 55–57, 63, 67]

Average amounts (ng/L)	Africa	Asia	Europe	Oceania	South America	References
ΣHCH	1135	4768	136.1		41.1	
HCH α	85.1	756.2	24.7		8.3	[2, 8, 10, 12–17, 19–23, 26, 27, 41, 48, 58, 71, 84, 85]
HCH β	91.5	1335	39.1		21.0	[2, 4, 10, 12–17, 19–21, 23, 26, 27, 41, 48, 58, 61, 85]
HCH δ	669.0	2299			4.5	[4, 14, 17, 19–21, 26, 85]
HCH γ	289.7	377.5	72.3		7.3	[2, 3, 7, 8, 13, 14, 16, 17, 19–23, 25–27, 30, 32, 34, 36, 38, 41, 47–50, 53–58, 63, 64, 66–68, 71, 84, 85, 93]
ΣHeptachlor, Heptachlor epoxide	580.6	31.1	34.5		1.6	
Heptachlor	150.6	24.1	17.7		0.9	[1, 2, 4, 8, 10, 12, 14, 16, 17, 19, 20, 23, 26, 41, 48, 53, 55–57, 59, 85]
Heptachlor epoxide	430.0	7.0	16.8		0.7	[8, 13, 14, 16, 17, 19, 20, 26, 41, 48, 50, 53, 57, 68, 85]
Heptachlor/ heptachlor epoxide	0.4	3.4	1.1		1.2	
Malathion	100.0	360.0	102.7		42.0	[6, 18, 40, 43, 45, 49, 50, 53, 56, 57, 63, 67, 72, 83, 92]
Methoxychlor	7.0	18.9	120.3			[4, 12–14, 16, 19, 35, 47, 50, 53, 56, 57]
Σ	7456	1054	8278	1875	419	

Data are displayed by category and continent referring to the most frequent pesticides (n ≥ 10). These values are based on the references cited in **Table 1**.

The pesticide names in bold are in the 2013/39/EU directive target list with specific MRLs; the ratio parent/residues is presented in italic style.

Table 3. Average values (ng/L) of the most frequent pesticides, quantified in water samples.

Another worth to mention aspect is that the concentrations of the herbicides, chlortoluron, and terbuthylazine in France exceeded ~300 and ~1900 ng/L, respectively, indicating an abusive application and/or improper waste treatment [34, 36]. However, none of these herbicides are included in the above referred European directive.

In Europe, the pesticides were highlighted in the Stockholm Convention list. Like DDT, aldrin, dieldrin, endrin, atrazine, HCB, HCH (gamma), heptachlor, heptachlor epoxide, mirex, and PeCB were quantified between 1996 and 2012. Average concentrations (**Figure 2**) ranged from 1.1 to 155 ng/L along these years, excluding 2004 when high concentrations of two cyclodiene pesticides (2377 ng/L for aldrin and 5156 ng/L for dieldrin) were registered in the same aquatic system (Lake Vela, Portugal) [51]. The parent/residues ratio for DDT and heptachlor reveals values above 1, indicating once again an active and abusive use of these pesticides.

Pesticides (ng/L)	BE	BG	FR	DE	GR	IT	NO	Pl	PT	RO	SP	NL	min-max
References					[29–40, 43, 45, 47–52, 54–71, 74, 75, 93]								
ΣDDT		3.7	180.5		65.8			14.1	126.7	3.0	23.6		3.0–180.5
4,4′-DDD			84.0					7.5	11.0	1.0	7.6		1.0–84.0
4,4′-DDE					30.8			0.0	19.7	1.0	3.6		0.02–30.8
4,4′-DDT		3.7	96.5		35.0			6.6	96.0	1.0	12.5		1.0–96.5
Alachlor			41.0	36.4	66.4				813.1		380.3		**36.4–813.1**
ΣCyclodienes		5.6			43.0			6.0	2174		6.4		5.6–2174.4
Aldrin		5.6			23.9			1.0	832.1		3.1		**1.0–832.1**
Dieldrin					19.2			5.0	1342		3.2		**3.2–1342**
Atrazine	213.7		95.6	18.6	67.9		0.1		459.7	77.0	30.6		0.1–459.7
Atz-desethyl			38.1	11.1	45.3				587.5		26.3		11.1–587.5
Chlorpyrifos					2.5	4.6			29.1		3.0		2.5–29.1
Chlortoluron			340.5	3.3					7.8		7.4	20.0	**3.3–340.5**
Diazinon					93.1	0.7			59.2	20.0	9.1		0.7–93.1
Dimethoate					5.2	2.3	0.0		11,650		23.7		**0.0–11,650**
Diuron	820.0		740.0	7.1			0.3		49.5		208.7		**0.3–820.0**
Endo. sulfate					19.1				55.2				19.1–55.2
Fenitrothion					3.3	2.3			77.9		151.5		2.3–151.5
HCH γ	56.0	12.8	200.0	1.3	25.7			83.4	146.6	2.6	15.7		1.3–200.0
Isoproturon	270.0		144.0	13.1			0.2		3.3		3.7	40.0	0.2–270.0
Malathion					19.4	2.7			41.8		229.1		2.7–229.1
Metolachlor	327.0		96.4	4.1			0.8		80.0		9.0		0.8–327.0
Simazine	71.9			7.7	2.7				33.5		156.9		2.7–156.9
Terbuthylazine	36.0		1950	4.1			0.3		78.9		414.0		**0.3–1950**
Terbutryn				203.4					37.4		3.7		3.7–203.4

Data are displayed by European country (BE: Belgium, BG: Bulgaria; FR: France; DE: Germany; GR: Greece; IT: Italy; NO: Norway; PL: Poland; PT: Portugal; RO: Romania; SP: Spain; NL: Netherlands) referring to the most frequent pesticides (n ≥ 10) based on the references cited in **Table 1**. Atz-desethyl: atrazine-desethyl; Endo. sulfate: endosulfan sulfate.
The pesticide names in bold are in the 2013/39/EU directive target list with specific MRLs.

Table 4. Average values (ng/L) of the most frequent pesticides quantified in water samples.

3. Final considerations

Water, as the most common analyzed matrix, is usually characterized by the quantification of pesticides dissolved in the aqueous phase (after filtering). In spite of its importance, more

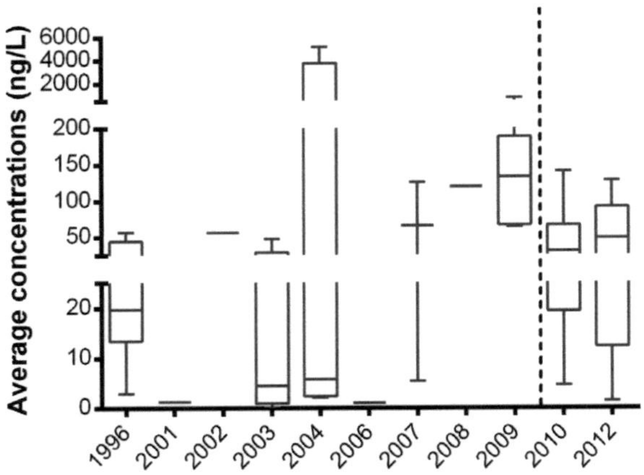

Figure 2. Representation of the priority listed pesticides average values (ng/L), quantified in water samples, collected in Europe, and displayed by sampling year (the dashed line separates the before and after the Stockholm Convention (2009).

efforts should be invested into quantifying pesticides present in the suspended particulate matter phase, since it is where most of the organic contaminants will be absorbed. In parallel, further legislative considerations should be applied. Looking at the number of different pesticides quantified per continent, Europe registered the highest number of compounds (141), which may be due to the amount of available data. Taking this in consideration together the category of the measured pesticides, insecticides were the most representative compounds, since they were measured in almost all continents, presenting also the highest number of cases above the European Legislative limits. This suggests that independently of the agricultural practices/needs, insecticides are the ones showing higher amounts in the aquatic systems. However, the highest average concentrations were registered in Asia, which can indicate an abusive usage of specific pesticides. Among continents, the continuous application of some pesticides scheduled for elimination in 2001 or 2009 by the Stockholm Convention is visible. As this study covers this transition time-frame, additional studies should be done to monitor the eradication of these substances.

In some cases, concentrations were clearly toxic to some trophic levels (acute concentrations); however, it is important to highlight that continuous exposure to medium/low levels (ng/L) may cause long-term adverse effects rippling into all trophic levels, in the likes of neurotoxicity, altered metabolism, endocrine disruption, and immunotoxicity in insects and invertebrates, passing through fish, amphibians, reptiles, and birds, and finally ending in mammals. Growth modulation, altered metabolism, and impaired photosynthesis may also occur in plants and fungi [91]. Further studies should also evaluate the impact of the main persistent metabolites, since they are the ones which persist longer in the aquatic systems.

In summary, further international discussions and pacts, such as the Stockholm Convention, should exist to alert mankind, to broadly regulate usages, monitoring, and where or when it is necessary to ban the use of these hazardous pesticides.

Acknowledgements

Partially supported by the European Regional Development Fund (ERDF) through the Operational Competitiveness Programme (COMPETE), and Operational Human Potential Programme (POPH), and by national Portuguese funds, through Foundation for Science and Technology (FCT), via project UID/Multi/04423/2013, project PTDC/MAR/70436/2006 (FCOMP-01-0124.FEDER-7382), grant (SFRH/BD/79305/2011), and grant (SFRH/BPD/120558/2016). This work was also implemented in the Framework of the Structured Program of R&D&I INNOVMAR—Innovation and Sustainability in the Management and Exploitation of Marine Resources (NORTE-01-0145-FEDER-000035), namely within the Research Line ECOSERVICES, supported by the Northern Regional Operational Programme (NORTE2020), through ERFD. Complementary funding was provided by ICBAS—U.Porto.

Author details

Catarina Cruzeiro[1,2,3], Eduardo Rocha[1,3] and Maria João Rocha[1,3]*

*Address all correspondence to: mjrocha@icbas.up.pt

1 Department of Microscopy, Laboratory of Histology and Embryology, ICBAS—Institute of Biomedical Sciences Abel Salazar, U.Porto—University of Porto, Porto, Portugal

2 Department of Life Sciences, CFE—Centre for Functional Ecology, University of Coimbra, Coimbra, Portugal

3 CIIMAR/CIMAR—Interdisciplinary Centre for Marine and Environmental Research, Group of Histomorphology, Pathophysiology and Applied Toxicology—Terminal de Cruzeiros do Porto de Leixões, Matosinhos, Portugal

References

[1] Agbohessi PT, Imorou Toko I, Ouédraogo A, Jauniaux T, Mandiki SNM, Kestemont P. Assessment of the health status of wild fish inhabiting a cotton basin heavily impacted by pesticides in Benin (West Africa). Science of the Total Environment. 2015;**506-507**:567-584

[2] Yamashita N, Urushigawa Y, Masunaga S, Walash MI, Miyazaki A. Organochlorine pesticides in water, sediment and fish from the Nile River and Manzala Lake in Egypt. International Journal of Environmental Analytical Chemistry. 2000;**77**(4):289-303

[3] Darko G, Akoto O, Oppong C. Persistent organochlorine pesticide residues in fish, sediments and water from Lake Bosomtwi, Ghana. Chemosphere. 2008;**72**(1):21-24

[4] Gbeddy G, Glover E, Doyi I, Frimpong S, Doamekpor L. Assessment of organochlorine pesticides in water, sediment, African cat fish and Nile tilapia, consumer exposure and

human health implications, Volta Lake. Ghana Journal of Environmental & Analitical Toxicology. 2015;**5**(4):297

[5] Otieno P, Okinda Owuor P, Lalah JO, Pfister G, Schramm KW. Monitoring the occurrence and distribution of selected organophosphates and carbamate pesticide residues in the ecosystem of Lake Naivasha, Kenya. Toxicological & Environmental Chemistry. 2015;**97**(1):51-61

[6] Sturve J, Scarlet P, Halling M, Kreuger J, Macia A. Environmental monitoring of pesticide exposure and effects on mangrove aquatic organisms of Mozambique. Marine Environmental Research. 2016;**121**:9-19

[7] Ize-Iyamu O, Asia I, Egwakhide P. Concentrations of residues from organochlorine pesticide in water and fish from some rivers in Edo State Nigeria. International Journal of Physical Sciences. 2007;**2**(9):237-241

[8] Ogbeide O, Tongo I, Ezemonye L. Risk assessment of agricultural pesticides in water, sediment, and fish from Owan River, Edo State, Nigeria. Environmental Monitoring and Assessment. 2015;**187**(10):1-16

[9] Ikpesu TO. Assessment of occurrence and concentrations of paraquat dichloride in water, sediments and fish from Warri River Basin, Niger Delta, Nigeria. Environmental Science and Pollution Research. 2015;**22**(11):8517-8525

[10] Awofolu R, Fatoki O. Persistent organochlorine pesticide residues in freshwater systems and sediments from the Eastern Cape, South Africa. Water SA. 2004;**29**(3):323-330

[11] Schulz R, Peall SKC, Dabrowski JM, Reinecke AJ. Current-use insecticides, phosphates and suspended solids in the Lourens River, Western Cape, during the first rainfall event of the wet season. Water SA. 2001;**27**(1):65-70

[12] Xue N, Xu X, Jin Z. Screening 31 endocrine-disrupting pesticides in water and surface sediment samples from Beijing Guanting reservoir. Chemosphere. 2005;**61**(11):1594-1606

[13] Shen B, Wu J, Zhao Z. Organochlorine pesticides and polycyclic aromatic hydrocarbons in water and sediment of the Bosten Lake, Northwest China. Journal of Arid Land. 2017;**9**(2):287-298

[14] Cui L, Ge J, Zhu Y, Yang Y, Wang J. Concentrations, bioaccumulation, and human health risk assessment of organochlorine pesticides and heavy metals in edible fish from Wuhan, China. Environmental Science and Pollution Research. 2015;**22**(20):15866-15879

[15] Yang D, Qi S, Zhang J, Wu C, Xing X. Organochlorine pesticides in soil, water and sediment along the Jinjiang River mainstream to Quanzhou Bay, southeast China. Ecotoxicology and Environmental Safety. 2013;**89**:59-65

[16] Zhang ZL, Hong HS, Zhou JL, Huang J, Yu G. Fate and assessment of persistent organic pollutants in water and sediment from Minjiang River Estuary, Southeast China. Chemosphere. 2003;**52**(9):1423-1430

[17] Zhou R, Zhu L, Yang K, Chen Y. Distribution of organochlorine pesticides in surface water and sediments from Qiantang River, East China. Journal of Hazardous Materials. 2006;**137**(1):68-75

[18] Chen H, Zhu J, Li Z, Chen A, Zhang Q. The occurrence and risk assessment of five organophosphorus pesticides in river water from Shangyu, China. Environmental Monitoring and Assessment. 2016;**188**(11):614

[19] Zhang Z, Huang J, Yu G, Hong H. Occurrence of PAHs, PCBs and organochlorine pesticides in the Tonghui River of Beijing, China. Environmental Pollution. 2004;**130**(2):249-261

[20] Tang Z, Yang Z, Shen Z, Niu J, Cai Y. Residues of organochlorine pesticides in water and suspended particulate matter from the Yangtze River catchment of Wuhan, China. Environmental Monitoring and Assessment. 2008;**137**(1-3):427-439

[21] Li J, Li F, Liu Q. Sources, concentrations and risk factors of organochlorine pesticides in soil, water and sediment in the Yellow River estuary. Marine Pollution Bulletin. 2015;**100**(1):516-522

[22] Lin T, Li J, Xu Y, Liu X, Luo C, Cheng H, et al. Organochlorine pesticides in seawater and the surrounding atmosphere of the marginal seas of China: Spatial distribution, sources and air–water exchange. Science of the Total Environment. 2012;**435-436**:244-252

[23] Huang Y, Xu Y, Li J, Xu W, Zhang G, Cheng Z, et al. Organochlorine pesticides in the atmosphere and surface water from the equatorial Indian Ocean: Enantiomeric signatures, sources, and fate. Environmental Science & Technology. 2013;**47**(23):13395-13403

[24] Singh S, Bhutia D, Sarkar S, Rai BK, Pal J, Bhattacharjee S, et al. Analyses of pesticide residues in water, sediment and fish tissue from river Deomoni flowing through the tea gardens of Terai Region of West Bengal, India. International Journal of Fisheries and Aquatic Studies. 2015;**3**:17-23

[25] Singare PU. Distribution and risk assessment of suspected endocrine-disrupting pesticides in creek water of Mumbai, India. Marine Pollution Bulletin. 2016;**102**(1):72-83

[26] Luo X, Mai B, Yang Q, Fu J, Sheng G, Wang Z. Polycyclic aromatic hydrocarbons (PAHs) and organochlorine pesticides in water columns from the Pearl River and the Macao harbor in the Pearl River Delta in South China. Marine Pollution Bulletin. 2004;**48**(11-12):1102-1115

[27] Carroll ML, Johnson BJ, Henkes GA, McMahon KW, Voronkov A, Ambrose WG Jr, et al. Bivalves as indicators of environmental variation and potential anthropogenic impacts in the southern Barents Sea. Marine Pollution Bulletin. 2009;**59**(4-7):193-206

[28] Chau NDG, Sebesvari Z, Amelung W, Renaud FG. Pesticide pollution of multiple drinking water sources in the Mekong Delta, Vietnam: Evidence from two provinces. Environmental Science and Pollution Research. 2015;**22**(12):9042-9058

[29] Loos R, Locoro G, Contini S. Occurrence of polar organic contaminants in the dissolved water phase of the Danube River and its major tributaries using SPE-LC-MS2 analysis. Water Research. 2010;**44**(7):2325-2335

[30] Direction générale des ressources naturelles et de l'environment. État des lieux des sous-bassins hydrographiques Tome I : état des lieux Sous-bassin Escaut—Lys. Incidences et évaluation du risque de non atteinte du bon état. Governmental document. Direction générale des Ressources naturelles et de l'Environment. 2005

[31] Noppe H, Ghekiere A, Verslycke T, Wulf ED, Verheyden K, Monteyne E, et al. Distribution and ecotoxicity of chlorotriazines in the Scheldt Estuary (B-Nl). Environmental Pollution. 2007;**147**(3):668-676

[32] Litskas VD, Dosis IG, Karamanlis XN, Kamarianos AP. Occurrence of priority organic pollutants in Strymon river catchment, Greece: Inland, transitional, and coastal waters. Environmental Science and Pollution Research. 2012;**19**(8):3556-3567

[33] Miège C, Schiavone S, Dabrin A, Coquery M, Mazzella N, Berho C, et al. An in situ inter-comparison exercise on passive samplers for monitoring metals, polycyclic aromatic hydrocarbons and pesticides in surface waters. TrAC Trends in Analytical Chemistry. 2012;**36**:128-143

[34] Pesce S, Fajon C, Bardot C, Bonnemoy F, Portelli C, Bohatier J. Longitudinal changes in microbial planktonic communities of a French river in relation to pesticide and nutrient inputs. Aquatic Toxicology. 2008;**86**(3):352-360

[35] Baugros J-B, Giroud B, Dessalces G, Grenier-Loustalot M-F, Cren-Olivé C. Multiresidue analytical methods for the ultra-trace quantification of 33 priority substances present in the list of REACH in real water samples. Analytica Chimica Acta. 2008;**607**(2):191-203

[36] Taghavi L, Merlina G, Probst J-L. The role of storm flows in concentration of pesticides associated with particulate and dissolved fractions as a threat to aquatic ecosystems—case study: The agricultural watershed of save river (Southwest of France). Knowledge and Management of Aquatic Ecosystems. 2011;(400):06

[37] Gasperi J, Garnaud S, Rocher V, Moilleron R. Priority substances in combined sewer overflows: Case study of the Paris sewer network. Water Science and Technology. 2011;**63**(5):853

[38] Koal T, Asperger A, Efer J, Engewald W. Simultaneous determination of a wide spectrum of pesticides in water by means of fast on-line SPE-HPLC-MS-MS—A novel approach. Chromatographia. 2003;**57**:S93-S101

[39] Quednow K, Püttmann W. Monitoring terbutryn pollution in small rivers of Hesse, Germany. Journal of Environmental Monitoring. 2007;**9**(12):1337-1343

[40] Thomatou A-A, Zacharias I, Hela D, Konstantinou I. Determination and risk assessment of pesticide residues in lake Amvrakia (W. Greece) after agricultural land use changes in the lake's drainage basin. International Journal of Environmental Analytical Chemistry. 2012;**93**(7):780-799

[41] Golfinopoulos SK, Nikolaou AD, Kostopoulou MN, Xilourgidis NK, Vagi MC, Lekkas DT. Organochlorine pesticides in the surface waters of Northern Greece. Chemosphere. 2003;**50**(4):507-516

[42] Lambropoulou DA, Sakkas VA, Hela DG, Albanis TA. Application of solid-phase micro-extraction in the monitoring of priority pesticides in the Kalamas River (N.W. Greece). Journal of Chromatography A. 2002;**963**(1-2):107-116

[43] Hela DG, Lambropoulou DA, Konstantinou IK, Albanis TA. Environmental monitoring and ecological risk assessment for pesticide contamination and effects in Lake Pamvotis, northwestern Greece. Environmental Toxicology and Chemistry. 2005;**24**(6):1548-1556

[44] Mörtl M, Németh G, Juracsek J, Darvas B, Kamp L, Rubio F, et al. Determination of glyphosate residues in Hungarian water samples by immunoassay. Microchemical Journal. 2013;**107**:143-151

[45] Montuori P, Aurino S, Nardone A, Cirillo T, Triassi M. Spatial distribution and partitioning of organophosphates pesticide in water and sediment from Sarno River and Estuary, Southern Italy. Environmental Science and Pollution Research. 2015;**22**(11):8629-8642

[46] Brumovský M, Bečanová J, Kohoutek J, Thomas H, Petersen W, Sørensen K, et al. Exploring the occurrence and distribution of contaminants of emerging concern through unmanned sampling from ships of opportunity in the North Sea. Journal of Marine Systems. 2016;**162**:47-56

[47] Badach H, Nazimek T, Kaminska IA. Pesticide content in drinking water samples collected from orchard areas in central Poland. Annals of Agricultural and Environmental Medicine. 2007;**14**(1):109

[48] Tomza-Marciniak A, Witczak A. Distribution of endocrine-disrupting pesticides in water and fish from the Oder River, Poland. Acta Ichthyologica Et Piscatoria. 2010;**40**(1):1-9

[49] Palma P, Kuster M, Alvarenga P, Palma VL, Fernandes RM, Soares AMVM, et al. Risk assessment of representative and priority pesticides, in surface water of the Alqueva reservoir (South of Portugal) using on-line solid phase extraction-liquid chromatography-tandem mass spectrometry. Environment International. 2009;**35**(3):545-551

[50] Rocha MJ, Ribeiro MFT, Cruzeiro C, Figueiredo F, Rocha E. Development and validation of a GC-MS method for determination of 39 common pesticides in estuarine water—Targeting hazardous amounts in the Douro River estuary. International Journal of Environmental Analytical Chemistry. 2012;**92**(14):1587-1608

[51] Abrantes N, Pereira R, Gonçalves F. Occurrence of pesticides in water, sediments, and fish tissues in a lake surrounded by agricultural lands: Concerning risks to humans and ecological receptors. Water, Air, & Soil Pollution. 2010;**212**(1):77-88

[52] AIASS A, Stigter TY. Multi-method assessment of nitrate and pesticide contamination in shallow alluvial groundwater as a function of hydrogeological setting and land use. Agricultural Water Management. 2009;**96**(12):1751-1765

[53] Cruzeiro C, Rocha E, Pardal MÂ, Rocha MJ. Environmental assessment of pesticides in the Mondego River Estuary (Portugal). Marine Pollution Bulletin. 2016;**103**:240-246

[54] Gonçalves CM, Esteves da Silva JCG, Alpendurada MF. Evaluation of the pesticide contamination of groundwater sampled over two years from a vulnerable zone in Portugal. Journal of Agricultural and Food Chemistry. 2007;**55**(15):6227-6235

[55] Cruzeiro C, Pardal M, Rocha E, Rocha M. Occurrence and seasonal loads of pesticides in surface water and suspended particulate matter from a wetland of worldwide interest—The Ria Formosa Lagoon, Portugal. Environmental Monitoring and Assessment. 2015;**187**(11):1-21

[56] Cruzeiro C, Rocha E, Pardal MÂ, Rocha MJ. Uncovering seasonal patterns of 56 pesticides in surface coastal waters of the Ria Formosa lagoon (Portugal), using a GC-MS method. International Journal of Environmental Analytical Chemistry. 2015;**95**(14):1370-1384

[57] Cruzeiro C, Rocha E, Pardal MÂ, Rocha MJ. Seasonal-spatial survey of pesticides in the most significant estuary of the Iberian Peninsula—The Tagus River Estuary. Journal of Cleaner Production. 2016;**126**:419-427

[58] Ferencz L, Balog AA. Pesticide survey in soil, water and foodstuffs from central Romania. Carpathian Journal of Earth and Environmental Sciences. 2010;**5**(1):111-118

[59] Dirtu D, Pancu M, Minea ML, Dirtu AC, Sandu I. Occurrence and assessment of selected chemical contaminants in drinking water from Eastern Romania. Revista die Chemie. 2016;**67**(10):2059-2064

[60] Köck-Schulmeyer M, Ginebreda A, de Alda M, Barceló D. Fate and Risks of Polar Pesticides in Groundwater Samples of Catalonia. The Handbook of Environmental Chemistry. Heidelberg: Springer Berlin; 2012. pp. 1-20

[61] Quintana J, Martí I, Ventura F. Monitoring of pesticides in drinking and related waters in NE Spain with a multiresidue SPE-GC-MS method including an estimation of the uncertainty of the analytical results. Journal of Chromatography A. 2001;**938**(1-2):3-13

[62] Pérez-Carrera E, León VML, Parra AG, González-Mazo E. Simultaneous determination of pesticides, polycyclic aromatic hydrocarbons and polychlorinated biphenyls in seawater and interstitial marine water samples, using stir bar sorptive extraction–thermal desorption–gas chromatography–mass spectrometry. Journal of Chromatography A. 2007;**1170**(1-2):82-90

[63] Postigo C, López de Alda MJ, Barceló D, Ginebreda A, Garrido T, Fraile J. Analysis and occurrence of selected medium to highly polar pesticides in groundwater of Catalonia (NE Spain): An approach based on on-line solid phase extraction–liquid chromatography–electrospray-tandem mass spectrometry detection. Journal of Hydrology. 2010;**383**(1-2):83-92

[64] Sánchez-Avila J, Tauler R, Lacorte S. Organic micropollutants in coastal waters from NW Mediterranean Sea: Sources distribution and potential risk. Environment International. 2012;**46**:50-62

[65] Hildebrandt A, Guillamón M, Lacorte S, Tauler R, Barceló D. Impact of pesticides used in agriculture and vineyards to surface and groundwater quality (North Spain). Water Research. 2008;**42**(13):3315-3326

[66] Claver A, Ormad P, Rodríguez L, Ovelleiro JL. Study of the presence of pesticides in surface waters in the Ebro river basin (Spain). Chemosphere. 2006;**64**(9):1437-1443

[67] Köck M, Farré M, Martínez E, Gajda-Schrantz K, Ginebreda A, Navarro A, et al. Integrated ecotoxicological and chemical approach for the assessment of pesticide pollution in the Ebro River delta (Spain). Journal of Hydrology. 2010;**383**(1-2):73-82

[68] Salvadó V, Quintana XD, Hidalgo M. Monitoring of nutrients, pesticides, and metals in waters, sediments, and fish of a wetland. Archives of Environmental Contamination and Toxicology. 2006;**51**(3):377-386

[69] Masiá A, Campo J, Vázquez-Roig P, Blasco C, Picó Y. Screening of currently used pesticides in water, sediments and biota of the Guadalquivir River basin (Spain). Journal of Hazardous Materials. 2013;**263**(Part 1):95-104

[70] Belmonte Vega A, Garrido Frenich A, Martínez Vidal JL. Monitoring of pesticides in agricultural water and soil samples from Andalusia by liquid chromatography coupled to mass spectrometry. Analytica Chimica Acta. 2005;**538**(1-2):117-127

[71] Menchen A, JDL H, JJG A. Pesticide contamination in groundwater bodies in the Júcar River European Union Pilot Basin (SE Spain). Environmental Monitoring and Assessment. 2017;**189**(4):146

[72] Köck-Schulmeyer M, Ginebreda A, González S, Cortina JL, de Alda ML, Barceló D. Analysis of the occurrence and risk assessment of polar pesticides in the Llobregat River Basin (NE Spain). Chemosphere. 2012;**86**(1):8-16

[73] Kuster M, López de Alda MJ, Hernando MD, Petrovic M, Martín-Alonso J, Barceló D. Analysis and occurrence of pharmaceuticals, estrogens, progestogens and polar pesticides in sewage treatment plant effluents, river water and drinking water in the Llobregat river basin (Barcelona, Spain). Journal of Hydrology. 2008;**358**(1-2):112-23

[74] Masiá A, Campo J, Navarro-Ortega A, Barceló D, Picó Y. Pesticide monitoring in the basin of Llobregat River (Catalonia, Spain) and comparison with historical data. Science of The Total Environment. 2015;**503-504**:58-68

[75] Moermond C, Vos J, Verbruggen E. Environmental Risk Limits for Organophosphorous Pesticides. Netherlands: National Institute for Public Health and the Environment (RIVM); 2008 Contract No.: 601714004

[76] Allinson G, Zhang P, Bui A, Allinson M, Rose G, Marshall S, et al. Pesticide and trace metal occurrence and aquatic benchmark exceedances in surface waters and sediments of urban wetlands and retention ponds in Melbourne, Australia. Environmental Science and Pollution Research. 2015;**22**(13):10214-10226

[77] Allinson G, Allinson M, Bui A, Zhang P, Croatto G, Wightwick A, et al. Pesticide and trace metals in surface waters and sediments of rivers entering the Corner Inlet Marine National Park, Victoria, Australia. Environmental Science and Pollution Research. 2016;**23**(6):5881-5891

[78] Mitchell C, Brodie J, White I. Sediments, nutrients and pesticide residues in event flow conditions in streams of the Mackay Whitsunday Region, Australia. Marine Pollution Bulletin. 2005;**51**(1-4):23-36

[79] Birch GF, Drage DS, Thompson K, Eaglesham G, Mueller JF. Emerging contaminants (pharmaceuticals, personal care products, a food additive and pesticides) in waters of Sydney estuary, Australia. Marine Pollution Bulletin. 2015;**97**(1-2):56-66

[80] Bainbridge ZT, Brodie JE, Faithful JW, Sydes DA, Lewis SE. Identifying the land-based sources of suspended sediments, nutrients and pesticides discharged to the Great Barrier Reef from the Tully–Murray Basin, Queensland, Australia. Marine and Freshwater Research. 2009;**60**(11):1081-1090

[81] De Gerónimo E, Aparicio VC, Bárbaro S, Portocarrero R, Jaime S, Costa JL. Presence of pesticides in surface water from four sub-basins in Argentina. Chemosphere. 2014;**107**:423-431

[82] Etchegoyen M, Ronco A, Almada P, Abelando M, Marino D. Occurrence and fate of pesticides in the Argentine stretch of the Paraguay-Paraná basin. Environmental Monitoring and Assessment. 2017;**189**(2):63

[83] Laabs V, Amelung W, Pinto AA, Wantzen M, da Silva CJ, Zech W. Pesticides in surface water, sediment, and rainfall of the northeastern Pantanal basin, Brazil. Journal of Environmental Quality. 2002;**31**(5):1636-1648

[84] Rissato SR, Galhiane MS, Ximenes VF, de Andrade RMB, Talamoni JLB, Libânio M, et al. Organochlorine pesticides and polychlorinated biphenyls in soil and water samples in the Northeastern part of São Paulo state, Brazil. Chemosphere. 2006;**65**(11):1949-1958

[85] Montory M, Ferrer J, Rivera D, Villouta MV, Grimalt JO. First report on organochlorine pesticides in water in a highly productive agro-industrial basin of the Central Valley, Chile. Chemosphere. 2017;**174**:148-156

[86] Production/crops [Internet]. FAOSTAT. 2015 [cited 28-12-2015]. Available from: http://faostat3.fao.org/

[87] Food and Agriculture Organization (FAO) of the United Nations. FAO Statistical Yearbook 2012—Africa Food and Agriculture. Accra, Ghana: FAO; 2012

[88] van den Berg L, Braun E, van der Meer J. Portugal: Urban policies or policies with an urban incidence? In: Limited AP, editor. National Policy Responses to Urban Challenges in Europe. England: Ashgate Publishing, Ltd.; 2007

[89] Branco MC, França FH, Medeiros MA, Leal JGT. Uso de inseticidas para o controle da traça-do-tomateiro e traça-das-crucíferas: um estudo de caso. Horticultura Brasileira. 2001;**19**:60-63

[90] Hunt L, Bonetto C, Resh VH, Buss DF, Fanelli S, Marrochi N, et al. Insecticide concentrations in stream sediments of soy production regions of South America. Science of the Total Environment. 2016;**547**:114-124

[91] Köhler H-R, Triebskorn R. Wildlife ecotoxicology of pesticides: Can we track effects to the population level and beyond? Science. 2013;**341**(6147):759-765

[92] Cruzeiro C, Pardal MÂ, Rodrigues-Oliveira N, Castro LFC, Rocha E, Rocha MJ. Multi-matrix quantification and risk assessment of pesticides in the longest river of the Iberian peninsula. Science of the Total Environment. 2016;**572**:263-272

[93] Pawełczyk A. Assessment of health risk associated with persistent organic pollutants in water. Environmental Monitoring and Assessment. 2013;**185**(1):497-508

[94] Executive Body of the Convention. Guidance Document on Best Available Techniques for Reducing Emissions of POPs from Major Stationary Sources Adopted on 18 December 2009. UNECE, 2009

[95] United Nations. Protocol to the 1979 convention on long-range transboundary air pollution on persistence organic pollutants—Amendments to the text and to annexes I, II, III, IV, VI and VIII to the protocol. 2009 Contract No.: C.N.555.2010.TREATIES-3

[96] Gervais G, Brosillon S, Laplanche A, Helen C. Ultra-pressure liquid chromatography-electrospray tandem mass spectrometry for multiresidue determination of pesticides in water. Journal of Chromatography A. 2008;**1202**(2):163-172

[97] Palma P, Köck-Schulmeyer M, Alvarenga P, Ledo L, López de Alda M, Barceló D. Environmental Science and Pollution Research. 2015;**22**(10):7665-7675

Chapter 4

Pesticides in Worldwide Aquatic Systems: Part II

Catarina Cruzeiro, Eduardo Rocha and
Maria João Rocha

Additional information is available at the end of the chapter

http://dx.doi.org/10.5772/intechopen.73117

Abstract

Contamination by pesticides is a worldwide problem that can greatly disturb the biota, directly and/or indirectly. Nonetheless, few efforts were done so far to present review-style publications that analyse and integrate monitoring data—in a global scale—and evaluate possible environmental risks. Herein, we assessed possible environmental risks through theoretical calculations, using worldwide data published at least during the last 17 years and considering different trophic levels and the maximum average environmental concentrations (in water) observed in each continent. Furthermore, hazard quotients—using the estimated average daily intake, theoretical maximum daily intake and the maximum residue limits—were calculated to estimate the potential risks to humans through direct consumption of molluscs, crustaceans and fish. In summary, several pesticides were quantified at concentrations capable to affect low to medium trophic level species, which through the food web can affect higher trophic levels; theoretical approaches considering the environmental mixtures showed that algae and invertebrates are the most sensitive groups. Moreover, fish and crustaceans evidenced the highest body concentrations. To evaluate a potential risk through direct consumption, human health risk assessments were done, and in spite of no direct risk, some hazard quotients indicate a potential risk for developing carcinogenic effects.

Keywords: insecticides, herbicides, fungicides, aquatic organisms, EC_{50}, LC_{50}, PNEC, ADI, EADI, MRL, hazard quotients, mollusc, crustacean, bivalve, fish, bioaccumulation, biomagnification

INTECH
open science | open minds

© 2018 The Author(s). Licensee InTech. This chapter is distributed under the terms of the Creative Commons Attribution License (http://creativecommons.org/licenses/by/3.0), which permits unrestricted use, distribution, and reproduction in any medium, provided the original work is properly cited. (cc) BY

1. Preamble

Worldwide, several studies have shown contamination with pesticides within different matrices. Together with the data shown previously in chapter "Pesticides in Worldwide Aquatic Systems: Part I", information such as the maximum concentrations in waters and the concentration of pesticides found in the different biological matrices were used to (i) assess eventual individual pesticide risk through the comparison with the well-established EC_{50}/LC_{50} for aquatic organisms, (ii) predict the environmental risk from pesticide mixtures found in each continent and (iii) assess the potential risk for human health when consuming molluscs, crustaceans and fish with the quantified concentrations.

2. Aquatic organisms

Fifty-two studies were used, where 111 different species were studied. The continent with the highest percentage of available results (quantified pesticides in different organisms) is Africa (39 species), followed by Europe (35 species), Asia (26 species) and then North and South America (nine and eight species, respectively). Here, we decided to focus on the sample type (zooplankton, molluscs, crustaceans, fish and mammals) analysed per continent and country (**Table 1**).

Continent/ country	Number of aquatic systems	Quantified pesticides	Sampling year	Sample type	av-min	av-max	av-av	References
					ng/g			
Africa								
Egypt	2	14	1993	C, F	1.1–6.3	7.6–8.2	4.1–130.2	[3]
Ethiopia	1	12	2011	F	4.1	27.2	17.1	[4]
Ghana	3	6–13	2004–2015	F	1.6–79.8	2.8–154.3	1.6–120.5	[4, 5]
Kenya	1	7	2011	F	na	na	0.3	[1, 6, 7]
Nigeria	5	1–65	2003–2014	F	19.5–3618	21.5–6355	20.6–5233	[1, 7–9]
Tunisia	1	14	2010	F	14.9	39.3	22.1	[10]
Asia								
China	5	7–45	2003–2013	F, Mo, C	0.6–2.5	1.8–34.5	0.8–11.8	[11–15]
India	1	3	na	Ma	na	na	74.5	[16]
Russia		8	2012–2013	F	18.1	52.1	31.7	[17]
South Korea	1	15	na	F	na	na	2.2	[18]
Tibet	1	55	2005	F	na	na	1.0	[19]
Europe								
Baltic Sea	1	18	2003	C, F, Mo	8.1	10.8	9.4	[20]
Belgium	1	5	2001	F, Mo	1.5	7.6	4.1	[21, 22]

Continent/ country	Number of aquatic systems	Quantified pesticides	Sampling year	Sample type	av-min	av-max	av-av	References
					ng/g			
Croatia	1	8	2012	Mo	0	1.4	na	[23]
Finland	1	5	2002	F	1.8	4.5	3.1	[24]
France	6	2–12	2001–2008	F, Mo	0.2–0.8	0.6–4.4	0.3–1.2	[21, 25, 26]
Italy	3	5–22	2002–2010	C, F, Mo	0.9–6.9	2.8–21.6	1.5–10.9	[10, 27, 28]
Poland	1	22	2003–2004	F	na	na	0.3	[29]
Portugal	4	1–54	2011–2013	F, Mo	4.6–7.6	27.2–72.0	0.2–18.6	[30–33]
Romania	1	72	2001	F,Z	188.3	278.4	220.3	[2]
Spain	3	3–13	1996–2015	C, F, Mo	0.1–10.4	0.8–9.0	0.3–8.0	[34–37]
North America								
California	1	19	2001	F	1.5	25.2	7.2	[38]
Canada	1	43	1999–2000	F	0.7	2.5	1.5	[39]
Greenland	1	18	1994–1995	F, Mo	0.1	0.3	0.2	[40]
Martinique Island	1	3	2003–2013	C, F, Mo	0.3	876.4	55.9	[41]
Mexico	1	15	2012–2013	F	2.1	25.5	6.3	[42]
USA	3	3–9	2004–2013	F, Mo	na	na	0.1–11.1	[43–45]
South America								
Argentina	1	6	1999	Ma	7	34.6	15.2	[46]
Brazil	3	6–42	1996–2009	F, Mo, Ma	0.1–30.1	0.1–410.7	0.1–99.8	[47–50]

Z, zooplankton; C, crustaceans; Mo, molluscs; F, fish; Ma, mammals; na, not applicable

Table 1. Pesticide concentrations [average minimum (av-min), average maximum (av-max) and average of averages (av-av) values; ng/g] fresh weight to make it; ng/g of fresh weight in aquatic organisms, presented by continent and country; the number of quantified aquatic systems, pesticides and sampling year were also added (when more than one aquatic system, a range of values are presented).

The data collected between 1993 and 2016 averaged from 0.1 to 5233 ng/g (**Table 1**). Europe is represented by 22 aquatic systems, followed by Africa, with 13, and the rest with no more than nine aquatic systems. When considering the number of pesticides quantified, Africa has more observations (382) than Europe (327) and the other continents (between 90 and 220), which is due to the higher number of species studied in Africa.

Africa stands out with average concentrations of 132 ng/g (SD 411), followed by Europe (57 ng/g, SD 271), South America (17 ng/g, SD 40) and Asia and North America (5 ng/g, SD 20). This scattered difference between concentrations is mainly due to the average values observed in Warri River (Nigeria, Africa) and in the Danube Delta (Romania, Europe) [1, 2].

Grouping data by category, insecticides prevail in 89% of biologic analyses, leaving 11% for the herbicide and fungicide categories and presenting the same pattern on all continents (**Figure 1**).

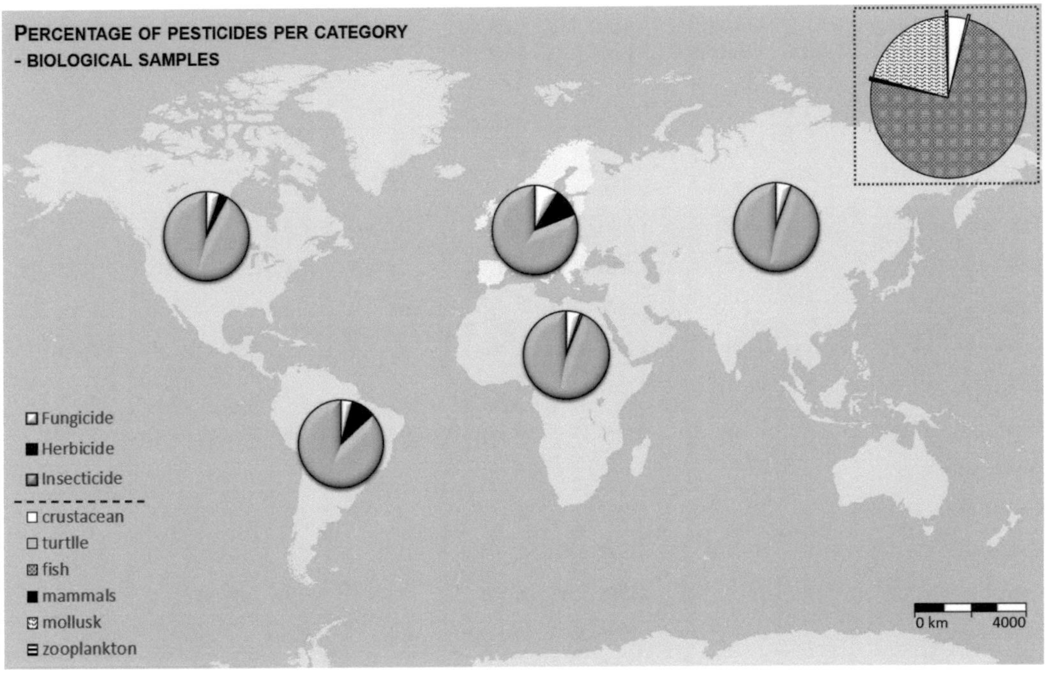

Figure 1. Representation of the quantified pesticides in organisms (%), per category, in each continent; the right upper corner pie chart represents the Metazoan lineages used worldwide.

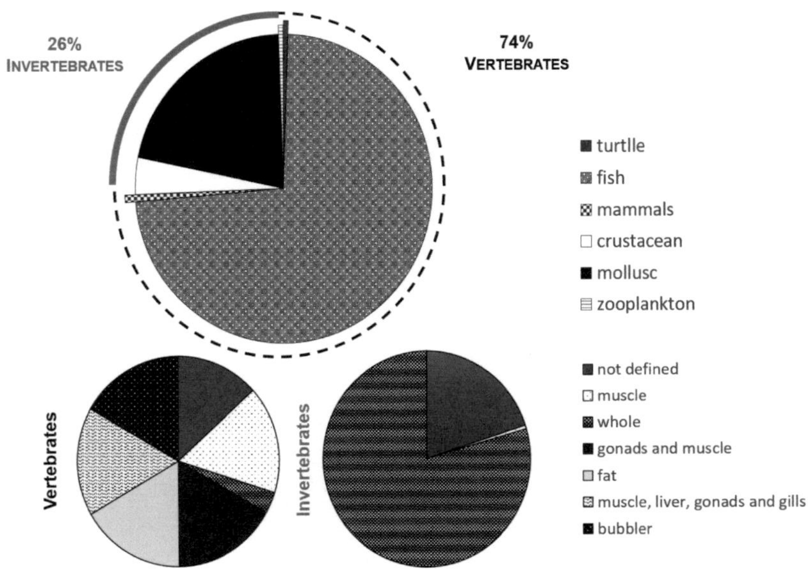

Figure 2. Representation of the quantified pesticides in organisms (%), per lineages of Metazoan, vertebrates and invertebrates and matrices.

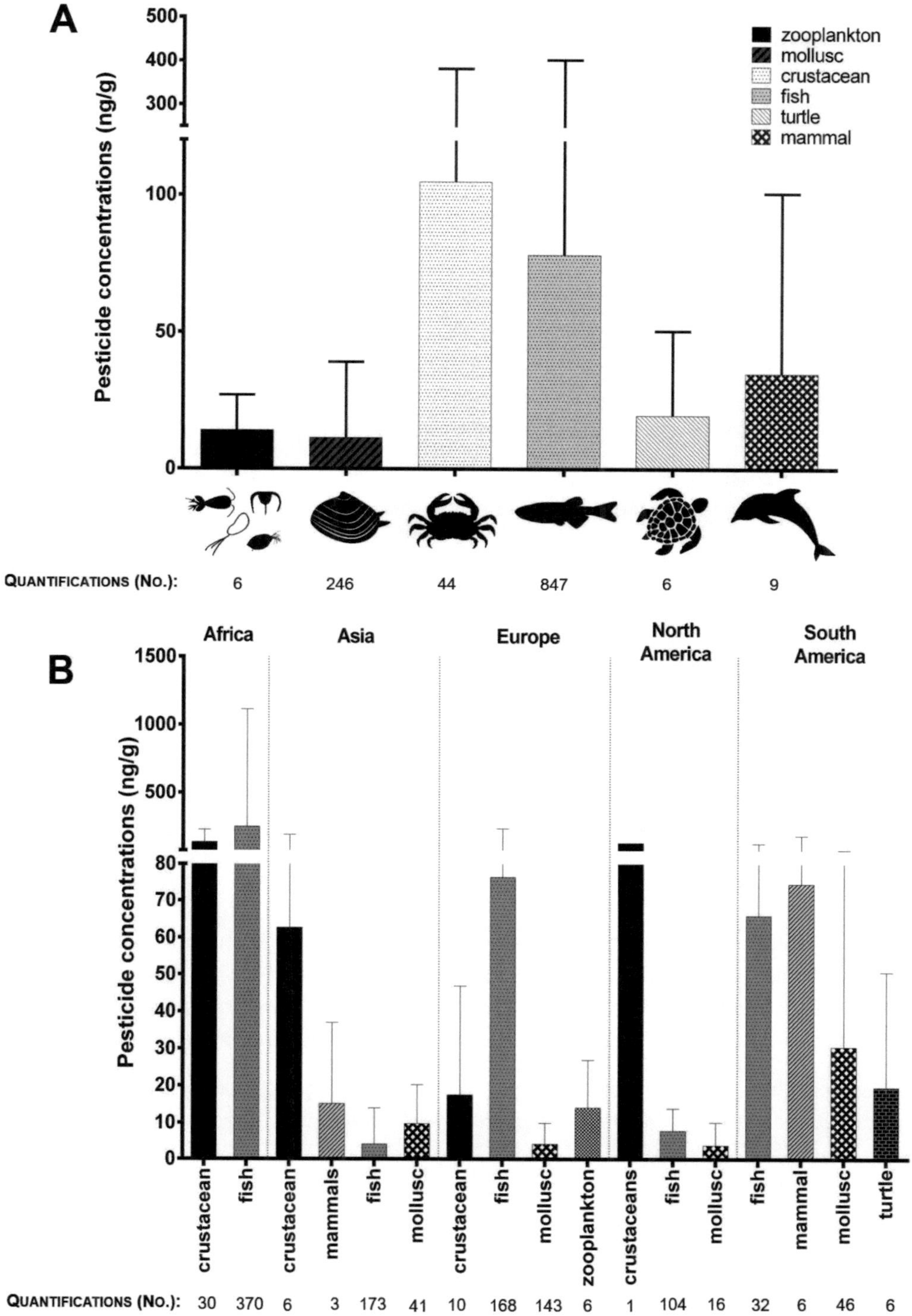

Figure 3. Average pesticide concentrations (ng/g fresh weigh) and number of quantifications per Metazoan lineage, worldwide (A) and by continent (B). The error bars represent standard deviations (SD).

No data are available for Oceania and Antarctica; so, when citing herein "worldwide", these continents will not appear.

Analysing data by matrix, the most common analysis occurred in fish (74%) and molluscs (20%). The remaining studies considered zooplankton, crustaceans, turtles and mammals (**Figure 1**). In total, 74% of the quantified pesticides were conducted on vertebrates and the other 26% in invertebrates (**Figure 1**). While for the latter, 80% of the quantifications were done using the whole animal, and for vertebrates, it is further divided; specific organs or tissues were used to quantify pesticides.

Many factors account for the strong bias towards vertebrates. Invertebrates are small, less complex and as a food resource almost entirely eatable, while the same is not applicable to vertebrates. Besides that, the study goal (i.e., food control or environmental/toxicological studies) also influences the type of the tissue/organ to study (muscle, liver, gonads or gills). For example, the bubbler tissue and fat are only applicable for aquatic mammals and turtles (**Figure 2**).

Results per Metazoan lineages (zooplankton, mollusc, crustacean, fish, turtles and mammals) were assessed considering the average concentrations and the number of quantifications (**Figure 3**). Average concentrations were ~11 ng/g for zooplankton and molluscs, ~35 ng/g for mammals and ~100 ng/g for crustaceans and fishes (**Figure 3A**).

Among continents, Africa presented the highest concentrations for crustaceans (142 ng/g) and fishes (253 ng/g) followed by North America (136 ng/g for crustaceans). Asia, Europe and South America included data belonging to four Metazoan lineages, with similar range of concentrations (~3 to ~76 ng/g (**Figure 3B**)).

3. Half effective and lethal concentrations (EC_{50}/LC_{50}) for aquatic organisms

It is now well established that at specific concentrations all pesticides are harmful to biota, affecting algae and plants, invertebrates and vertebrates [51]. Databases such as Pesticides Properties DataBase (PPDB) present information on the physicochemical properties, environmental fate, human health and ecotoxicological data of all active ingredients and approved pesticides [52].

In order to evaluate the worst-case scenario, the maximum average concentrations measured in waters from each continent were compared against the acute and chronic concentrations for aquatic animals, documented by the PPDB (see chapter "Part I: Pesticides in Worldwide Aquatic Systems"). On a global scale, 57 pesticides were registered at maximum average concentrations above the LC_{50} and/or EC_{50} settled for algae, invertebrates and/or fishes; among continents, Europe reported the highest number of pesticides (44 of 116), followed by Asia (14 of 42), Africa, Oceania and finally South America (6 of 24).

The most critical measured environmental concentrations (MEC) were registered for dicofol, ethion (Asia), metribuzin (Europe) and diazinon (Africa) with values from 2- to 200-folds higher than EC_{50} or LC_{50} set for invertebrates and algae.

4. Predictive aquatic risk assessment of pesticide mixtures

Despite the common occurrence of pesticides, mixtures in the environment, laws, conventions and recommendations still focus on individual standard parameters. Modelling approaches, based on available ecotoxicological information, can be used to estimate the impact of mixtures in the biota, completing this lack of information [53].

Based on the European chemical legislation REACH, the ecological risk quotient (RQ) is determined by the equation:

$$\mathbf{RQ}\left(\frac{\mathbf{MEC}}{\mathbf{PNEC}}\right) = \frac{\text{Measured Environmental Concentration (MEC; mg/L)}}{\text{Predicted No Effect Concentration (PNEC; mg/L)}}$$

PNEC is derived by selecting the most sensitive trophic level—from algae, invertebrate or fish—and applying an appropriate assessment factor (AF) [52, 54]. The AF, also denoted as safety or uncertainty factor, considers intra- and inter-laboratory variation of the data, biological variance and short-term to long-term exposures, presenting stipulated values for specific conditions [55, 56]; as an example, considering the Maximum Acceptable Concentration-Quality Standards (MAC-QS) to assess short-term effects an AF = 100 should be applied [55].

The RQ values, classified from <0.01 (negligible) to >1 (significant), indicate a range of potential risks for concern, but do not inform about the specific biological end point for that organism which is representing a specific trophic level [53, 57]. For this reason, a second approach, which defines the most sensitive trophic level for that environmental concentration, should be applied [53]:

$$\mathbf{RQ\ toxic\ units\ (TU)} = \frac{\text{MEC (mg/L)}}{EC_{50}\text{or } LC_{50} \text{ per each trophic level (mg/L)}}$$

RQ_{TU} values are summed per trophic level (sum of the toxic units (RQ_{STU})). If both $RQ_{(MEC/PNEC)}$ and RQ_{STU} are >1, additional considerations are required [53]. Based on the two reference models—concentration addition (CA) and independent action (IA)—the RQ_{STU}/Max_{TU} can be used to predict the second tier, resulting in the maximum value from which CA may display higher toxicity values than IA [58].

In this work, the maximum of the average measured concentration of pesticides in water samples was used to assess the potential risk per continent and on a worldwide scale (**Table 2**). From a total of 144 pesticides quantified in water samples, 133 were used for ecological risk assessment (**Table 2**); the remainders, mostly isomers and metabolites, were not integrated due to lack of information on their EC_{50} and LC_{50} concentrations set for these trophic levels (algae, invertebrate and fish). The highest number of pesticides suitable for this approach are represented by insecticides (n = 118). In general, algae was the most sensitive group to herbicides and fungicides, with 75% and 61.5% of the cases, respectively, while invertebrates showed the highest sensitivity to insecticides (66.1%) (**Table 2**).

Globally, the $RQ_{(MEC/PNEC)}$ resulted in 43% of very high-risk cases, led by insecticides; fungicides were the least worrisome category, as most of the cases presented low or negligible risks (**Figure 4**).

	Africa	Asia	Europe	Oceania	South America	PNEC	Algae	Invert	Fish
			mg/L			(mg/L)		%	
Herbicides									
Av.	5.7E-04	8.1E-05	7.8E-03	5.7E-04	2.3E-03	2.5E-06–2.0E+00	75.0	6.25	18.8
n	5	6	38	9	7				
Insecticides									
Av.	7.7E-04	2.2E-03	2.3E-03	1.4E-05	2.9E-04	1.7E-08–1.0E+00	8.5	25.4	66.1
n	19	28	49	7	15				
Fungicides									
Av.	8.5E-05	4.5E-04	2.6E-04	1.1E-05	3.9E-05	3.0E-05–4.6E-01	61.5	0.0	38.5
n	1	8	22	3	3				

Table 2. Ecological risk assessment through the PNEC, using the maximum average concentrations of pesticides in water (mg/L), quantified in each continent; here in this table only the average values/category (Av.), the total number of pesticides (n) observed per category and continent, and the range of PNEC values are presented; data based on **Table 1** of the chapter in this book *entitled* Pesticides in worldwide aquatic systems- Part I.

The results presented above are a consequence of the highest values measured around the world. Since Europe was the continent with more values of $RQ_{(MEC/PNEC)}$, these results are mostly representative for this continent (see **Table 1**). However, this does not mean that concentrations measured on the other continents are innocuous. As observed for the number of compounds analysed per continent, Africa presented the most disturbing scenarios (52%), followed by Asia and Europe (45%) and then Oceania and South America (24%) with RQ > 1 (**Figure 5**).

In order to evaluate the effect of the maximum average concentrations found per individual trophic level (RQ_{TU}), further evaluation should be done through RQ_{STU} (**Table 3**).

When comparing between continents, the highest RQ_{STU} ratios were attained in Europe, for algae (16.13) and fish (33.12), and in Asia for invertebrates (324.97); however, the last one is due to a punctual concentration observed in India for ethion [59]. Independently of that, the invertebrate group is the most sensitive trophic level, presenting the highest RQ_{STU} values. The same pattern is observed in the other continents except in Oceania, where the highest risk is observed for the algae (0.92) by the herbicides (**Table 3**).

The $RQ_{(MEC/PNEC)}$ and RQ_{STU} demonstrate that one or more biotest organisms are sensitive to the concentrations presented on that continent; so, the ratio RQ_{STU}/highest RQ_{TU} was done, applying the highest sum among trophic levels (**Table 4**).

For each of these scenarios, the maximal possible ratio RQ_{STU}/RQ_{TU} was lower than the value given by the number of mixture of toxic components, suggesting that the possible observed toxicity is due to a low number of pesticides. As we can notice, the RQ_{STU}/RQ_{TU} ratio is very

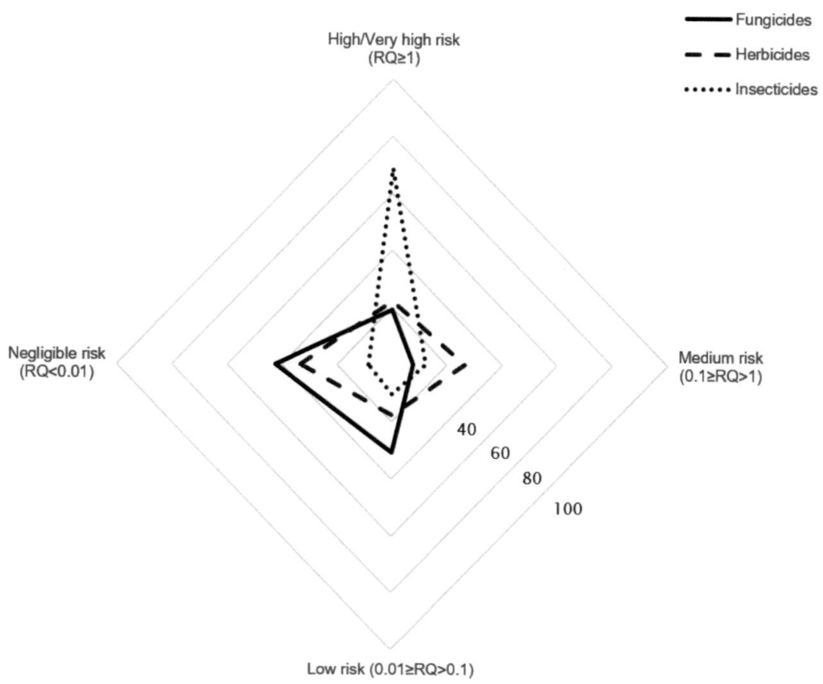

Figure 4. Distribution of pesticides per category (%), according to $RQ_{(MEC/PNEC)}$ ranking.

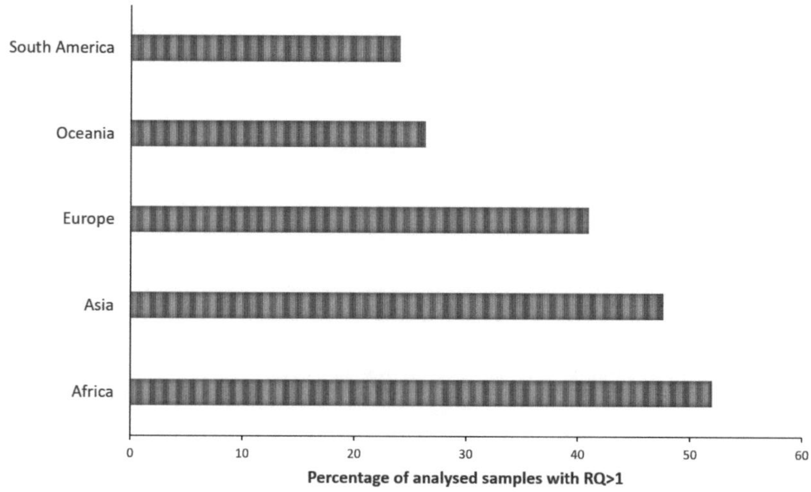

Figure 5. Percentage of $RQ_{(MEC/PNEC)}$ samples above 1, grouped by continent (total number of observations, n= 25, 42, 110, 19, 25 in Africa, Asia, Europe, Oceania, and South America, respectively).

	RQ$_{STU}$				
	Africa	Asia	Europe	Oceania	South America
Algae					
Fungicides	0.01	0.07	0.03	0.00	0.00
Herbicides	0.08	0.00	15.28	0.91	0.01
Insecticides	0.04	0.27	0.81	0.00	0.01
Total	0.13	0.34	16.13	**0.92**	0.02
Invertebrates					
Fungicides	0.00	0.06	0.02	0.00	0.00
Herbicides	0.57	2.91	0.19	0.00	0.02
Insecticides	8.65	322.0	18.84	0.01	3.94
Total	**9.22**	**324.97**	19.05	0.01	**3.96**
Fishes					
Fungicides	0.00	0.01	0.02	0.00	0.00
Herbicides	0.17	0.02	0.59	0.00	0.00
Insecticides	0.67	8.42	32.51	0.00	0.43
Total	0.84	8.45	**33.12**	0.00	0.43

Table 3. Sum of the toxic units per trophic level (RQ$_{STU}$) of each continent (with available data), organised by pesticide category; the most sensitive trophic level, per continent, is in bold.

Continent	No. of compounds (toxic/total)	RQ$_{TU}$		ΣRQ$_{STU}$			ΣRQ$_{STU}$/RQ$_{TU}$
				Algae	Invert	Fishes	
Africa	13/25	Parathion methyl	8.05		9.22		1.15
Asia	10/21	Ethion	221.42	8.45			1.47
Europe	9/22	Deltamethrin	20.11	33.12			1.65
Oceania	5/19	Diuron	0.79			0.92	1.17
South America	6/25	Cypermethrin	2.47		3.96		1.61

Table 4. Second tier, using RQ$_{STU}$ and the highest RQ$_{TU}$ per trophic level and continent.

similar among continents; however, the number of toxic compounds per total, where Africa presents a significant number (52%) when compared to the others, should be also considered (**Table 4**).

5. Human health risks

Dietary pesticide risks can be estimated through well-established indices and defined and used by US Environmental Protection Agency (EPA) [60], European Food Safety Authority (EFSA) and European Union Directives [61, 62]. Realistic predictions involve several parameters, such as pesticide residue intake (PRI), as the one reported by Food and Agriculture Organisation (FAO) [63]:

$$\mathbf{PRI} = \text{Pesticide Concentration (mg/kg)} \times \text{Acceptable Consumption Rate} \left(\text{kg/capita/day}\right)$$

The acceptable daily intake (ADI) estimates the amount of a substance in food that can be ingested daily over a lifetime without appreciable health risk to the consumer [64]:

$$\mathbf{ADI(mg/kg/day)} = \frac{\text{No observed Effect Level (NOEL)}}{\text{Safety Factor}}$$

The estimated average daily intake (EADI), according to EPA, should be less than the established ADI values [64]:

$$\mathbf{EADI}\left(\frac{\mathbf{mg\ bw}}{\mathbf{kg\ day}}\right) = \frac{\text{PRI}}{\text{Standard Body Weight}}$$

The theoretical maximum daily intake (TMDI) represents the maximum concentration of a pesticide residue (mg/kg) legally permitted in food [64]:

$$\mathbf{TMDI} = \text{Comsumption Rate} \left(\text{kg/capita/day}\right) \times \text{Maximum Residue Limits (MRLs)}$$

When no specific MRL is published, a 0.01 mg/kg value is applied [8]. Additionally, hazard quotients (HQs)—which measure the potential exposure for developing non-carcinogenic health effects—may be calculated using several assumptions [65].

EADI may be divided by the acute reference dose (ARfD, mg/kg/day) [14]—which is derived from the no-observed-adverse-effect levels (NOAEL) and based on studies of short time exposures (1–7 days) [66], by ADI, for long intake periods, and by TMDI, which is advised by EFSA to calculate the potential risks of unintentional compounds, such as pollutants.

In the chapter Part I: Pesticides in Worldwide Aquatic Systems, the levels/categories of pesticides per continent/country are displayed. The maximum average concentrations shown in *Part I* were used here to assess human health risks. Data are summarised in **Table 5**.

Continent	Molluscs	Invertebrates	Fishes
Africa	—	1.6E + 00 (30)	5.2E + 00 (370)
Asia	3.2E-01 (41)	1.0E-02 (3)	9.0E-02 (173)
Europe	6.0E-02 (143)	1.4E-01 (10)	4.0E + 00 (168)
North America	1.0E-02 (16)	1.4E-01 (1)	3.0E-02 (104)
South America	1.8E-01 (46)	—	1.7E-01 (32)

Table 5. Average maximum concentrations (mg/kg) found per continent and by group of aquatic animals (mollusc, invertebrates and fishes) and the total number of cases used in each case (between brackets).

	Continent/pesticide	MEC	EADI	MRL	TMDI	ADI	ARfD	HQ			
								EADI/ MRL	EADI/ TMDI	EADI/ ADI	EADI/ ARfD
Mollusc	**Asia**										
	HCH (gamma)	2.9E-03	2.9E-06	2.0E-02	1.1E-03	8.0E-03	3.0E-04	0	0.02	0.08	0
	HCH (sigma)	3.2E-01	3.1E-04	1.0E-02	5.7E-04	—	—	0.03	0.09	0.21	—
	Heptachlor epoxide	5.8E-04	5.8E-07	4.0E-03	2.3E-04	1.0E-04	—	0	0.07	0.08	—
	Methoxychlor	2.2E-03	2.2E-06	1.0E-02	5.7E-04	1.0E-01	5.0E-03	0	0.55	—	—
	Europe										
	Cyanazine	3.6E-02	3.6E-05	1.0E-02	5.8E-04	2.0E-03	—	0	0.04	0.09	—
	Endrin	1.8E-02	1.8E-05	5.0E-02	2.9E-03	2.0E-04	3.0E-04	0	0.06	0.14	—
	HCH (gamma)	7.7E-03	7.7E-06	2.0E-02	1.2E-03	8.0E-03	3.0E-04	0	0.05	0.16	0.01
	Parathion ethyl	9.2E-03	9.2E-06	5.0E-02	2.9E-03	6.0E-04	5.0E-03	0	0.1	0.06	0
	Phosmet	4.2E-02	4.1E-05	1.0E-01	5.8E-03	1.0E-02	4.5E-02	0	0.05	0.15	0
	Procymidone	1.5E-02	1.5E-05	1.0E-02	5.8E-04	2.8E-03	1.2E-02	0	0.06	0	0
	Propazine	1.4E-02	1.4E-05	1.0E-02	5.8E-04	2.0E-02	1.7E-02	0	0.06	0.02	—
	Propyzamide	1.2E-02	1.2E-05	1.0E-02	5.8E-04	2.0E-02	7.5E-02	0	0.07	0.01	0
	Simetryn	1.3E-02	1.2E-05	1.0E-02	5.8E-04	2.5E-02	—	0	0.08	0	0
	Terbuthylazine	2.4E-02	2.4E-05	1.0E-02	5.8E-04	4.0E-03	8.0E-03	0	0.08	0	0
	Terbutryn	9.5E-03	9.5E-06	1.0E-02	5.8E-04	1.0E-01	1.0E-03	0	0.09	0.05	0.01
	Tetrachlorvinphos	4.5E-02	4.5E-05	1.0E-02	5.8E-04	5.0E-02	3.0E-02	0	0.11	—	0.01
	South America										
	Mirex	2.6E-05	8.3E-09	1.0E-02	2.5E-04	—	2.0E-04	0	0.53	0.54	—
	Pentachlorobenzene	8.3E-05	2.6E-08	1.0E-02	2.5E-04	1.7E-02	—	0	0.11	0.01	0.19

	Continent/pesticide	MEC	EADI	MRL	TMDI	ADI	ARfD	HQ			
								EADI/MRL	EADI/TMDI	EADI/ADI	EADI/ARfD
Crustacean	**Africa**										
	∑DDD,DDE,DDT	2.5E+00	1.2E-03	1.0E +00	2.9E-02	5.0E-03	—	0	0.04	0.24	—
	Chlordane (alpha)	3.0E-02	1.5E-05	2.0E-03	5.9E-05	5.0E-04	—	0.01	0.25	0.03	—
	Chlordane (gamma)	5.5E-01	2.6E-04	1.0E-02	2.9E-04	5.0E-04	—	0.03	0.9	0.53	—
	Hexachlorobenzene	3.6E-02	1.7E-05	1.0E-02	2.9E-04	6.0E-04	8.0E-04	0	0.06	0.03	0.02
	Nonachlor (beta)	1.9E-02	9.2E-06	6.0E-03	1.8E-04	—	—	0	0.05	—	—
	North America										
	Chlordecone	1.4E-01	9.9E-05	1.0E-02	5.8E-04	—	—	0.01	0.17	—	—
Fish	**Africa**										
	∑Aldrin + dieldrin	1.2E+00	5.6E-04	6.0E-03	1.8E-04	1.0E-04	3.0E-03	0.09	3.19	5.62	0.19
	∑DDD,DDE,DDT	3.5E+00	1.7E-03	1.0E +00	2.9E-02	5.0E-03	—	0	0.06	0.34	—
	Atrazine	6.3E-01	3.1E-04	1.0E-02	2.9E-04	2.0E-02	1.0E-01	0.03	1.04	0.02	0
	Carbofuran	2.2E-01	1.1E-04	2.0E-02	5.9E-04	1.5E-04	1.5E-04	0.01	0.18	0.71	0.71
	Chlordane (alpha)	1.8E-01	8.7E-05	2.0E-03	5.9E-05	5.0E-04	—	0.04	1.48	0.17	—
	Chlordane (gamma)	1.2E-01	5.8E-05	1.0E-02	2.9E-04	5.0E-04	—	0.01	0.2	0.12	—
	ΣEndosulfan	8.6E-01	4.2E-04	5.0E-02	1.5E-03	6.0E-03	2.0E-02	0.01	0.28	0.07	0.02
	Endrin	4.5E-01	2.2E-04	5.0E-02	1.5E-03	2.0E-04	3.0E-04	0	0.15	1.09	0.73
	Endrin aldehyde	3.3E+00	1.6E-03	1.0E-02	2.9E-04	2.0E-04	3.0E-04	0.16	5.37	7.89	5.26
	HCH (alpha)	1.9E+00	9.0E-04	2.0E-01	5.9E-03	—	—	0	0.15	—	—

Continent/pesticide	MEC	EADI	MRL	TMDI	ADI	ARfD	HQ			
							EADI/ MRL	EADI/ TMDI	EADI/ ADI	EADI/ ARfD
HCH (beta)	2.3E+00	1.1E-03	1.0E-01	2.9E-03	—	—	0.01	0.39	—	—
HCH (gamma)	6.6E-01	3.2E-04	2.0E-02	5.9E-04	8.0E-03	3.0E-04	0.02	0.54	0.04	1.07
Heptachlor	7.4E-01	3.6E-04	4.0E-03	1.2E-04	1.0E-04	—	0.09	3.05	3.58	—
Heptachlor epoxide	2.5E-01	1.2E-04	4.0E-03	1.2E-04	1.0E-04	—	0.03	1.03	1.21	—
Hexachlorobenzene	5.4E-01	2.6E-04	1.0E-02	2.9E-04	6.0E-04	8.0E-04	0.03	0.89	0.44	0.33
Nonachlor (beta)	5.2E-02	2.5E-05	6.0E-03	1.8E-04	—	—	0	0.14	—	—
Paraquat dichloride	5.2E+00	2.5E-03	1.0E-02	2.9E-04	4.0E-03	4.0E-04	0.25	8.62	0.63	6.34
Europe										
∑Aldrin + dieldrin	1.0E-02	1.0E-05	6.0E-03	3.5E-04	1.0E-04	3.0E-03	0	0.03	0.1	0
∑DDD,DDE,DDT	5.5E+00	5.5E-03	1.0E +00	5.8E-02	5.0E-03	—	0.01	0.1	1.11	—
HCH (gamma)	6.0E-02	6.0E-05	2.0E-02	1.2E-03	8.0E-03	3.0E-04	0	0.05	0.01	0.2
Hexachlorobenzene	5.7E-02	5.7E-05	1.0E-02	5.8E-04	6.0E-04	8.0E-04	0.01	0.1	0.09	0.07
North America										
Chlordane	9.6E-03	6.9E-06	2.0E-03	1.2E-04	5.0E-04	—	0	0.06	0.01	—
Chlordane (alpha)	1.0E-02	7.2E-06	2.0E-03	1.2E-04	5.0E-04	—	0	0.06	0.01	—
Endrin	1.6E-02	1.2E-05	5.0E-02	2.9E-03	2.0E-04	3.0E-04	0	0	0.06	0.04
Heptachlor	7.7E-03	5.6E-06	4.0E-03	2.3E-04	1.0E-04	—	0	0.02	0.06	—
South America										
∑Aldrin + dieldrin	1.5E-01	4.7E-05	6.0E-03	1.5E-04	1.0E-04	3.0E-03	0.01	0.31	0.47	0.02

http://dx.doi.org/10.5772/intechopen.73117

Continent/pesticide	MEC	EADI	MRL	TMDI	ADI	ARfD	HQ			
							EADI/	EADI/	EADI/	EADI/
							MRL	TMDI	ADI	ARfD
Endrin aldehyde	5.2E-02	1.6E-05	1.0E-02	2.5E-04	2.0E-04	3.0E-04	0	0.06	0.08	0.05
HCH (gamma)	1.6E-01	5.1E-05	2.0E-02	5.1E-04	8.0E-03	3.0E-04	0	0.1	0.01	0.17
Heptachlor	3.3E-02	1.1E-05	4.0E-03	1.0E-04	1.0E-04	—	0	0.1	0.11	—
Heptachlor epoxide	1.8E-02	5.7E-06	4.0E-03	1.0E-04	1.0E-04	—	0	0.06	0.06	—

MEC, measured environmental concentration (mg/kg); **EADI**, estimated average daily intake (mg/kg bw); **MRL**, maximum residue limit (mg/kg); **TMDI**, theoretical maximum daily intake (mg); **ADI**, acceptable daily intake (mg/kg bw/d); **ARfD**, acute reference dose (mg/kg bw/day); **fish and seafood consumption** (kg/capita/day): 0.0294 (Africa), 0.05705 (Asia), 0.05765 (Europe), 0.05833 (North America), 0.02548 (South America); **body weight** (kg): 60.7 (Africa), 57.7 (Asia), 70.8 (Europe), 80.7 (North America) and 67.9 (South America).

Table 6. Human health hazard, associated with mollusc, crustaceans and fish consumption, displayed by continent and pesticide.

The highest concentrations were observed in fish from Africa (5.2 mg/kg) and Europe (4.0 mg/kg), followed then by crustaceans in Africa (1.6 mg/kg). The highest number of cases (number of quantifications found considering all the pesticides, species and countries) was registered for fish (169 average cases), followed by molluscs (62 average cases), and finally crustaceans (11 average cases). The elevated number of fish studies is likely due to their importance as a food source.

For allowing a detailed evaluation of human health hazard, the same data is displayed by pesticide and continent and organised considering molluscs, crustaceans and fishes (**Table 6**). The food consumption rate and the average adult body weight were defined by continent [63, 64]. For the compounds endrin ketone and aldehyde, HCH (sigma and lambda), pretilachlor and pentachlorobenzene, a MRL of 0.01 mg/kg was adopted, since no specific data was found.

Focusing on the molluscs results, the MEC of 15, 52, 10 and 16 pesticides (from Asia, Europe, North America and South America, respectively) were used to calculate the HQs. Due to the low ratio values, only cases with at least one ratio value above 0.05 were presented. As we can see, none of the results proved to be harmful to human through direct consumption. In other words, none of the ratios were above 1, indicating that the calculated EADI was below the reference levels (MRL, TMDI, ADI and ARfD). The highest $HQ_{(EADI/TMDI)}$ was obtained for methoxychlor in Asia (0.55). For $HQ_{(EADI/ADI)}$, the highest ratio occurred in South America for mirex with 0.54.

Looking to the crustacean data, a total MEC of eight, two, three and one cases from Africa, Asia, Europe, North America, respectively, were analysed. The same criterion, which is the case with at least one ratio value above 0.05, was applied. High HQs for chlordane (gamma) were observed in crustaceans sampled in Africa (see **Table 6**). In spite of that, none of the ratios were above 1.

Twenty-four (Africa), 16 (Asia), 10 (Europe), 28 (North America) and 21 (South America) MEC cases were analysed considering the fish data. Once again, only HQ ratios with at least one case above 0.05 are shown. As we can see, none of the maximum average concentrations were above the MRL values; however, several HQ > 1 are observed in Africa, bringing potential exposure for developing carcinogenic health effects. This fact may be a result of bioaccumulation processes (where concentrations increase in higher trophic levels) and/or a higher interest in this matrix (increasing the data availability and diversity). These ratios were registered for six compounds—∑aldrin + dieldrin, endrin aldehyde, paraquat dichloride, endrin, heptachlor and heptachlor epoxide—where the most preoccupant cases (HQ > 3) are for the first three pesticides cited above.

6. Final considerations

Globally, and because of these high average concentrations, several individual pesticides were quantified at levels exceeding the established LC_{50} for fish and EC_{50} for invertebrates and algae.

In addition, the review has provided clear evidence that the biological data grouped according to Metazoan lineages reached higher concentrations for fish and crustaceans (**Figure 3**). It is worth noting however that the same pattern was not verified for higher trophic levels including turtles and aquatic mammals which may be due to the lack of samples. Considering that globally, many of the data displayed a wide range of concentrations, coupled with the fact that many of the larger aquatic species are migratory; there is a need to address the pesticide problem from a global perspective.

As a complement to this work, all edible species were evaluated for dietary pesticide risks, as mollusc, crustaceans and fish. No direct human health risk was observed; however, in Africa, some hazard quotients (HQ) were above one, indicating a potential exposure for developing carcinogenic health effects.

In conclusion, the potentially harmful effects of pesticides should be considered not only locally (national/governmental institutions) but also on a global scale.

Acknowledgements

This study was partially supported by the European Regional Development Fund (ERDF) through the Operational Competitiveness Programme (COMPETE) and Operational Human Potential Programme (POPH) and by national Portuguese funds, through Foundation for Science and Technology (FCT), via project UID/Multi/04423/2013 and project PTDC/MAR/70436/2006 (FCOMP-01-0124.FEDER-7382) and grant (SFRH/BD/79305/2011) and grant (SFRH/BPD/120558/2016). This work was also implemented in the framework of the Structured Program of R&D&I INNOVMAR (Innovation and Sustainability in the Management and Exploitation of Marine Resources; Reference NORTE-01-0145-FEDER-000035), namely, within the research line ECOSERVICES, supported by the Northern Regional Operational Programme (NORTE2020), through ERFD. Complementary funding was provided by ICBAS-U.Porto.

Author details

Catarina Cruzeiro[1,2,3], Eduardo Rocha[1,3]* and Maria João Rocha[1,3]

*Address all correspondence to: mjrocha@icbas.up.pt

1 ICBAS—Institute of Biomedical Sciences Abel Salazar, Department of Microscopy, Laboratory of Histology and Embryology, U.Porto—University of Porto, Porto, Portugal

2 CFE—Centre for Functional Ecology, Department of Life Sciences—University of Coimbra, Coimbra, Portugal

3 CIIMAR/CIMAR—Interdisciplinary Centre for Marine and Environmental Research, Group of Histomorphology, Pathophysiology and Applied Toxicology—Terminal de Cruzeiros do Porto de Leixões, Matosinhos, Portugal

References

[1] Ikpesu TO. Assessment of occurrence and concentrations of paraquat dichloride in water, sediments and fish from Warri River basin, Niger Delta, Nigeria. Environmental Science and Pollution Research. 2015;**22**(11):8517-8525

[2] Covaci A, Gheorghe A, Hulea O, Schepens P. Levels and distribution of organochlorine pesticides, polychlorinated biphenyls and polybrominated diphenyl ethers in sediments and biota from the Danube Delta, Romania. Environmental Pollution. 2006;**140**(1):136-149

[3] Yamashita N, Urushigawa Y, Masunaga S, Walash MI, Miyazaki A. Organochlorine pesticides in water, sediment and fish from the Nile River and Manzala Lake in Egypt. International Journal of Environmental Analytical Chemistry. 2000;**77**(4):289-303

[4] Norli HR, Christiansen A, Deribe E. Application of QuEChERS method for extraction of selected persistent organic pollutants in fish tissue and analysis by gas chromatography mass spectrometry. Journal of Chromatography A. 2011;**1218**(41):7234-7241

[5] Akoto O, Andoh H, Darko G, Eshun K, Osei-Fosu P. Health risk assessment of pesticides residue in maize and cowpea from Ejura, Ghana. Chemosphere. 2013;**92**

[6] Omwenga I, Kanja L, Nguta J, Mbaria J, Irungu P, Ng CA. Organochlorine pesticide residues in farmed fish in Machakos and Kiambu counties, Kenya. Cogent environmental. Science. 2016;**2**(1):1153215

[7] Ize-Iyamu O, Asia I, Egwakhide P. Concentrations of residues from organochlorine pesticide in water and fish from some rivers in Edo state Nigeria. International journal of physics. Sciences. 2007;**2**(9):237-241

[8] Ogbeide O, Tongo I, Ezemonye L. Risk assessment of agricultural pesticides in water, sediment, and fish from Owan River, Edo state, Nigeria. Environmental Monitoring and Assessment. 2015;**187**(10):1-16

[9] Yehouenou A. Pazou E, Lalèyè P, Boko M, van Gestel CAM, Ahissou H, Akpona S, et al. Contamination of fish by organochlorine pesticide residues in the Ouémé River catchment in the republic of Bénin. Environment International. 2006;**32**(5):594-599

[10] Ameur WB, Trabelsi S, El Megdiche Y, Ben Hassine S, Barhoumi B, Hammami B, et al. Concentration of polychlorinated biphenyls and organochlorine pesticides in mullet (*Mugil cephalus*) and sea bass (*Dicentrarchus labrax*) from Bizerte lagoon (northern Tunisia). Chemosphere. 2012;**90**(9):2372-2380

[11] Cui L, Ge J, Zhu Y, Yang Y, Wang J. Concentrations, bioaccumulation, and human health risk assessment of organochlorine pesticides and heavy metals in edible fish from Wuhan, China. Environmental Science and Pollution Research. 2015;**22**(20):15866-15879

[12] Jin MQ, Zhou SS, Liu WP, Zhang D, Lu XT. Residues and potential health risks of DDTs and HCHs in commercial seafoods from two coastal cities near Yangtze River estuary. Journal of Environmental Science and Health, Part B. 2015;**50**(3):163-174

http://dx.doi.org/10.5772/intechopen.73117

[13] Jiang QT, Lee TKM, Chen K, Wong HL, Zheng JS, Giesy JP, et al. Human health risk assessment of organochlorines associated with fish consumption in a coastal city in China. Environmental Pollution. 2005;**136**(1):155-165

[14] Zhao Z, Zhang L, Wu J, Fan C. Distribution and bioaccumulation of organochlorine pesticides in surface sediments and benthic organisms from Taihu Lake, China. Chemosphere. 2009;**77**(9):1191-1198

[15] Yang Y, Liu M, Xu S, Hou L, Ou D, Liu H, et al. HCHs and DDTs in sediment-dwelling animals from the Yangtze estuary, China. Chemosphere. 2006;**62**(3):381-389

[16] Adeleye AO, Jin H, Di Y, Li D, Chen J, Ye Y. Distribution and ecological risk of organic pollutants in the sediments and seafood of Yangtze estuary and Hangzhou Bay, East China Sea. Science of The Total Environment. 2016;**541**:1540-1548

[17] Tsygankov VY, Lukyanova ON, Khristoforova NK. The sea of Okhotsk and the Bering Sea as the region of natural aquaculture: Organochlorine pesticides in Pacific salmon. Marine Pollution Bulletin. 2016;**113**(1–2):69-74

[18] Choi M, Lee I-S, Jung R-H. Rapid determination of organochlorine pesticides in fish using selective pressurized liquid extraction and gas chromatography–mass spectrometry. Food Chemistry. 2016;**205**:1-8

[19] Yang R, Yao T, Xu B, Jiang G, Xin X. Accumulation features of organochlorine pesticides and heavy metals in fish from high mountain lakes and Lhasa River in the Tibetan plateau. Environment International. 2007;**33**(2):151-156

[20] Galassi S, Bettinetti R, Neri MC, Jeannot R, Dagnac T, Bristeau S, et al. A multispecies approach for monitoring persistent toxic substances in the Gulf of Gdańsk (Baltic sea). Ecotoxicology and Environmental Safety. 2008;**69**(1):39-48

[21] Schnitzler JG, Thomé JP, Lepage M, Das K. Organochlorine pesticides, polychlorinated biphenyls and trace elements in wild European sea bass (*Dicentrarchus labrax*) off European estuaries. Science of the Total Environment. 2011;**409**(19):3680-3686

[22] Wille K. Analytical Approaches for Quantification of Emerging Micropollutants in the Belgian Coastal Zone Gent: Gent University; 2011

[23] Milun V, Lušić J, Despalatović M. Polychlorinated biphenyls, organochlorine pesticides and trace metals in cultured and harvested bivalves from the eastern Adriatic coast (Croatia). Chemosphere. 2016;**153**:18-27

[24] Pikkarainen A-L. Polychlorinated biphenyls and organochlorine pesticides in Baltic Sea sediments and bivalves. Chemosphere. 2007;**68**(1):17-24

[25] Thomas M, Lazartigues A, Banas D, Brun-Bellut J, Feidt C. Organochlorine pesticides and polychlorinated biphenyls in sediments and fish from freshwater cultured fish ponds in different agricultural contexts in north-eastern France. Ecotoxicology and Environmental Safety. 2012;**77**(0):35-44

[26] Luna-Acosta A, Budzinski H, Le Menach K, Thomas-Guyon H, Bustamante P. Persistent organic pollutants in a marine bivalve on the Marennes-Oléron Bay and the Gironde estuary (French Atlantic Coast)—Part 1: Bioaccumulation. Science of The Total Environment. 2015;**514**:500-510

[27] Ferrante MC, Clausi MT, Meli R, Fusco G, Naccari C, Lucisano A. Polychlorinated biphenyls and organochlorine pesticides in European eel (*Anguilla anguilla*) from the Garigliano River (Campania region, Italy). Chemosphere. 2010;**78**(6):709-716

[28] Perugini M, Cavaliere M, Giammarino A, Mazzone P, Olivieri V, Amorena M. Levels of polychlorinated biphenyls and organochlorine pesticides in some edible marine organisms from the central Adriatic Sea. Chemosphere. 2004;**57**(5):391-400

[29] Tomza-Marciniak A, Witczak A. Distribution of endocrine-disrupting pesticides in water and fish from the Oder River, Poland. Acta Ichthyologica Et Piscatoria. 2010;**40**(1):1-9

[30] Abrantes N, Pereira R, Gonçalves F. Occurrence of pesticides in water, sediments, and fish tissues in a lake surrounded by agricultural lands: Concerning risks to humans and ecological receptors. Water, Air, & Soil Pollution. 2010;**212**(1):77-88

[31] Cruzeiro C, Pardal MÂ, Rodrigues-Oliveira N, Castro LFC, Rocha E, Rocha MJ. Multimatrix quantification and risk assessment of pesticides in the longest river of the Iberian peninsula. Science of The Total Environment. 2016;**572**:263-272

[32] Grilo TF, Cardoso PG, Pato P, Duarte AC, Pardal MA. Organochlorine accumulation on a highly consumed bivalve (*Scrobicularia plana*) and its main implications for human health. Science of The Total Environment. 2013;**461–462**:188-197

[33] Cruzeiro C, Rodrigues-Oliveira N, Velhote S, Pardal MÂ, Rocha E, Rocha MJ. Development and application of a QuEChERS–based extraction method for the analysis of 55 pesticides in the bivalve *Scrobicularia plana* by GC–MS/MS. Analytical and Bionalytical Chemistry. 2016;**408**:3681-3698

[34] Salvadó V, Quintana XD, Hidalgo M. Monitoring of nutrients, pesticides, and metals in waters, sediments, and fish of a wetland. Archives of Environmental Contamination and Toxicology. 2006;**51**(3):377-386

[35] Rodríguez-Hernández Á, Camacho M, Henríquez-Hernández LA, Boada LD, Valerón PF, Zaccaroni A, et al. Comparative study of the intake of toxic persistent and semi persistent pollutants through the consumption of fish and seafood from two modes of production (wild-caught and farmed). Science of The Total Environment. 2017;**575**:919-931

[36] Sánchez-Avila J, Fernandez-Sanjuan M, Vicente J, Lacorte S. Development of a multiresidue method for the determination of organic micropollutants in water, sediment and mussels using gas chromatography–tandem mass spectrometry. Journal of Chromatography A. 2011;**1218**(38):6799-6811

[37] Suárez P, Ruiz Y, Alonso A, San Juan F. Organochlorine compounds in mussels cultured in the Ría of Vigo: Accumulation and origin. Chemosphere. 2013;**90**(1):7-19

[38] Sapozhnikova Y, Bawardi O, Schlenk D. Pesticides and PCBs in sediments and fish from the Salton Sea, California, USA. Chemosphere. 2004;**55**(6):797-809

[39] Easton MDL, Luszniak D, Von der Geest E. Preliminary examination of contaminant loadings in farmed salmon, wild salmon and commercial salmon feed. Chemosphere. 2002;**46**(7):1053-1074

[40] Cleemann M, Riget F, Paulsen GB, Klungsøyr J, Dietz R. Organochlorines in Greenland marine fish, mussels and sediments. Science of the Total Environment. 2000;**245**(1–3):87-102

[41] Dromard CR, Bodiguel X, Lemoine S, Bouchon-Navaro Y, Reynal L, Thouard E, et al. Assessment of the contamination of marine fauna by chlordecone in Guadeloupe and Martinique (lesser Antilles). Environmental Science and Pollution Research. 2016;**23**(1): 73-80

[42] Granados-Galván IA, Rodríguez-Meza DG, Luna-González A, González-Ocampo HA. Human health risk assessment of pesticide residues in snappers (Lutjanus) fish from the Navachiste lagoon complex, Mexico. Marine Pollution Bulletin. 2015;**97**(1–2):178-187

[43] Nilsen EB, Hapke WB, McIlraith B, Markovchick D. Reconnaissance of contaminants in larval Pacific lamprey (*Entosphenus tridentatus*) tissues and habitats in the Columbia River basin, Oregon and Washington, USA. Environmental Pollution. 2015;**201**:121-130

[44] Granek EF, Conn KE, Nilsen EB, Pillsbury L, Strecker AL, Rumrill SS, et al. Spatial and temporal variability of contaminants within estuarine sediments and native Olympia oysters: A contrast between a developed and an undeveloped estuary. Science of The Total Environment. 2016;**557–558**:869-879

[45] Blocksom KA, Walters DM, Jicha TM, Lazorchak JM, Angradi TR, Bolgrien DW. Persistent organic pollutants in fish tissue in the mid-continental great rivers of the United States. Science of The Total Environment. 2010;**408**(5):1180-1189

[46] Durante CA, Neto EBS, Azevedo A, Crespo EA, Lailson-Brito JPOP. In the south Latin America: Bioaccumulation of DDT, PCB, HCB, HCH and Mirex in blubber of common dolphin (*Delphinus delphis*) and Fraser's dolphin (*Lagenodelphis hosei*) from Argentina. Science of the Total Environment. 2016;**572**:352-360

[47] Sánchez-Sarmiento AM, Rossi S, Vilca FZ, Thijl Vanstreels RE, Monteiro SH, Vale LAS, et al. Organochlorine pesticides in green sea turtles (*Chelonia mydas*) with and without fibropapillomatosis caught at three feeding areas off Brazil. Journal of the Marine Biological Association of the United Kingdom. 2016;**97**(1):215-223

[48] DMLd S, PBd C, Martinelli LA, Lanças FM, Pinto JS, Avelar WEP. Organochlorine pesticides in Piracicaba River basin (São Paulo/Brazil): A survey of sediment, bivalve and fish. Química Nova. 2008;**31**(2):214-219

[49] Miranda AL, Roche H, Randi MAF, Menezes ML, Ribeiro CAO. Bioaccumulation of chlorinated pesticides and PCBs in the tropical freshwater fish Hoplias Malabaricus: Histopathological, physiological, and immunological findings. Environment International. 2008;**34**(7):939-949

[50] Galvao P, Henkelmann B, Longo R, Lailson-Brito J, Torres JPM, Schramm K-W, et al. Distinct bioaccumulation profile of pesticides and dioxin-like compounds by mollusk bivalves reared in polluted and unpolluted tropical bays: Consumption risk and seasonal effect. Food Chemistry. 2012;**134**(4):2040-2048

[51] Köhler H-R, Triebskorn R. Wildlife ecotoxicology of pesticides: Can we track effects to the population level and beyond? Science. 2013;**341**(6147):759-765

[52] The Pesticide Properties DataBase (PPDB) developed by the Agriculture & Environment Research Unit (AERU) [Internet]. funded by UK national sources and through EU-funded projects. 2013 [cited 22–01-2013]. Available from: http://sitem.herts.ac.uk/aeru/footprint/index2.htm

[53] Backhaus T, Faust M. Predictive environmental risk assessment of chemical mixtures: a conceptual framework. Environmental Science & Technology. 2012;**46**(5):2564-2573

[54] Compound summary-toxicity [Internet]. U.S. National Library of Medicine. 2015. Available from: https://pubchem.ncbi.nlm.nih.gov/

[55] Proposal for a directive of the European parliament and of the council amending directives 2000/60/EC and 2008/105/EC as regards priority substances in the field of water policy (2012)

[56] Malkiewicz K, Hansson SO, Rudén C. Assessment factors for extrapolation from short-time to chronic exposure—Are the REACH guidelines adequate? Toxicology Letters. 2009;**190**(1):16-22

[57] Sánchez-Bayo F, Baskaran S, Kennedy IR. Ecological relative risk (EcoRR): Another approach for risk assessment of pesticides in agriculture. Agriculture, Ecosystems & Environment. 2002;**91**(1–3):37-57

[58] Silva E, Cerejeira M. Concentration addition-based approach for aquatic risk assessment of realistic pesticide mixtures in Portuguese river basins. Environmental Science and Pollution Research. 2014:1-10

[59] Singh S, Bhutia D, Sarkar S, Rai BK, Pal J, Bhattacharjee S, et al. Analyses of pesticide residues in water, sediment and fish tissue from river Deomoni flowing through the tea gardens of Terai region of West Bengal, India. International Journal of Fisheries and Aquatic Studies. 2015;**3**:17-23

[60] US Environmental Protection Agency (EPA). Guidelines for Ecological Risk Assessment. Washington, US: EPA; 1998

[61] Directive 2008/105/EC of the European parliament and of the council of 16 December 2008 on environmental quality standards in the field of water policy, amending and subsequently repealing council directives 82/176/EEC, 83/513/EEC, 84/156/EEC, 84/491/EEC, 86/280/EEC and amending directive 2000/60/EC of the European parliament and of the council, L 348/34, 2008

[62] European Food Safety Authority (EFSA). Scientific report of EFSA- The 2011 European Union report on pesticide residues in food. Parma, Italy: Journal 2014; 2014. p. 511

[63] Food balance/food supply - livestock and fish primary equivalent [Internet]. FAO. 2015 [cited 3–11-2015]. Available from: http://faostat3.fao.org/

[64] Regulation (EC) N 396/2005 of the European Parliament and of the Council of 23 February 2005 on maximum residue levels of pesticides in or on food and feed of plant and animal origin and amending Council Directive 91/414/EEC, 2005

[65] Herrman JL, Younes M. Background to the ADI/TDI/PTWI. Regulatory Toxicology and Pharmacology. 1999;**30**(2):S109-SS13

[66] Walpole SC, Prieto-Merino D, Edwards P, Cleland J, Stevens G, Roberts I. The weight of nations: An estimation of adult human biomass. BMC Public Health. 2012;**12**(1):1-6

The Ecology and Food Web Dynamics of South African Intermittently Open Estuaries

Pierre William Froneman

Additional information is available at the end of the chapter

http://dx.doi.org/10.5772/intechopen.72859

Abstract

This chapter provides an overview of the ecology and food web dynamics of southern African intermittently open/closed estuaries (IOCEs). Intermittently open/closed estuaries experience periodic isolation from the ocean due to a sandbar at the mouth and account for some 71% of all estuaries along the southern African coastline. Field studies indicate that the ecosystem functioning of IOCEs is strongly linked mouth phase (open vs. closed) of these systems. During the closed phase, these systems are generally characterised by low biological diversity and elevated biomass of both invertebrates and vertebrates, which are thought to be sustained by elevated biomass of microphytoplankton and zooplankton within these systems. The low diversity can be related to the virtual absence of marine species within these systems due to the presence of a sandbar at the mouth which limits recruitment. The overflow of marine waters into the estuary during winter storms or spring high tides contributes to the recruit of marine breeding species into these systems. Heavy rainfall in the catchment areas of these systems culminates in the water levels of these systems rising until such time the estuary breaches. The breaching event coincides with the outflow of biologically rich estuarine waters into the marine environment and provides an opportunity for marine breeding species to recruit into these systems. Global warming is likely to contribute to changes in the hydrodynamics of these systems with a concurrent impact on the food webs of these systems.

Keywords: southern Africa, estuaries, intermittently open/closed, ecology, global change

1. Introduction

The South African coastline stretching some 3100 km can broadly be divided into three biogeographic zones: the warm subtropical zone along the east coast, the warm temperate

INTECH
open science | open minds

© 2018 The Author(s). Licensee InTech. This chapter is distributed under the terms of the Creative Commons Attribution License (http://creativecommons.org/licenses/by/3.0), which permits unrestricted use, distribution, and reproduction in any medium, provided the original work is properly cited.

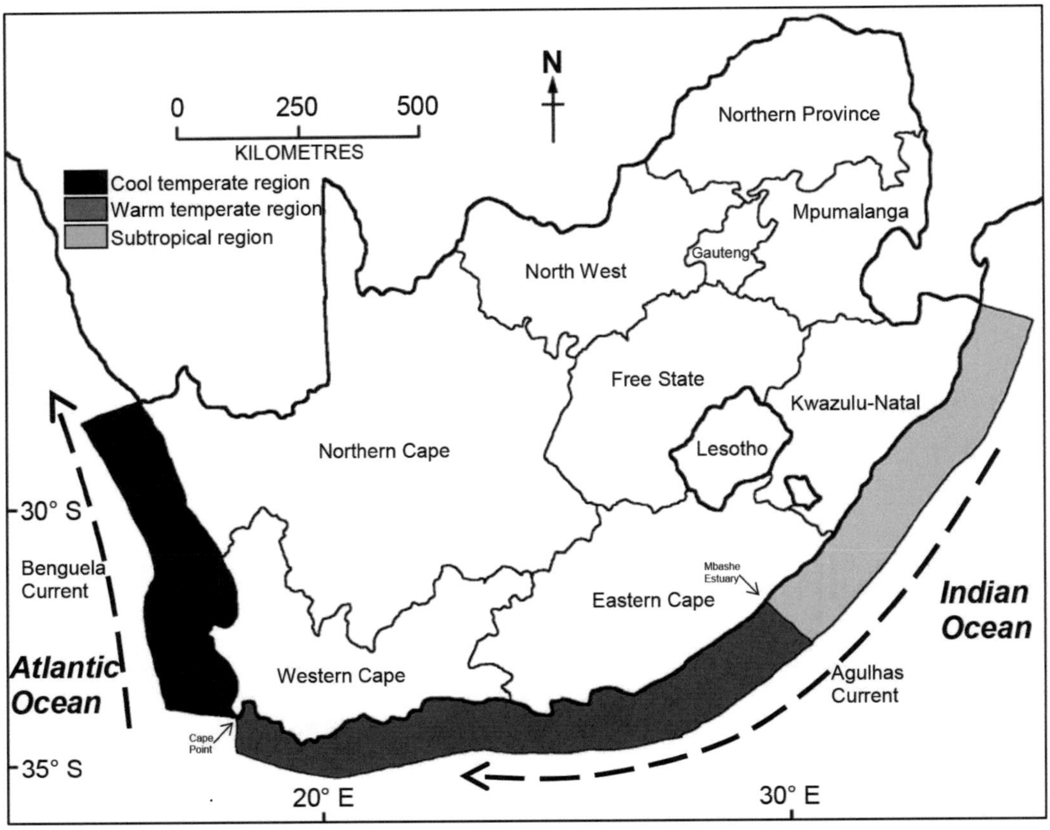

Figure 1. The geographic extent of the three biogeographical zones along the South African coastline (after Whitfield 1992b).

zone along the south coast and the cool temperature zone along the west coast (**Figure 1**). Within these three zones, there are 258 functional estuaries of which 71% can be categorised as intermittently open/closed (IOCEs) or temporary open/closed estuaries [1, 2]. Intermittently open/closed systems experience periodic isolation from the ocean, usually during periods of drought or no river inflow, during which a sand berm develops across the mouth of the estuary [2, 3]. Following periods of high rainfall (>100 mm in a month) and freshwater run-off, the volume of water in the estuary rises until it exceeds the height of the sandbar [1–3]. It is at this stage that breaching usually occurs, with a consequent drastic drop in water level exposing large areas of substratum [2, 4]. During the subsequent period, the estuary will be tidally dominated until such time that long-shore drift contributes to the reformation of the sandbar at the mouth of the system.

A link to the marine environment can also be established through the overtopping of marine waters across the sand bank at the mouth of the estuary during winter storms or during spring high tides [2, 3]. In addition to altering the physico-chemical properties (temperature, salinity and dissolved oxygen) in the lower reaches of these systems, these events represent important recruitment opportunities for marine breeding invertebrates and vertebrates, mainly ichthyofauna into these systems [5, 6].

There are two main types of ICOEs found along the southern African coastline, non-perched estuaries, which predominate in the subtropical zone and perched estuaries which are found along the east and west coast of the country [2, 7] (**Figure 2**). Perched estuaries have an average surface water level exceeding that of the marine environment [2, 7]. Non-perched systems are the more common system with the average surface water level similar to that of the marine environment [7]. Generally, lower salinities are observed in perched systems due to the

Figure 2. Images of an intermittently open/closed estuary under open conditions (a–c), closed conditions (d–f) and the mouth of the estuary under open conditions (g–i). Photographs courtesy of G Tweddle.

reduced frequency of occurrence over washing events with the dominant source of water being freshwater run-off [7]. Moreover, when a perched system breaches, it drains rapidly and will not experience tidal ebbing due to its height above the marine high water mark [7, 8]. Non-perched systems are tidally dominated following a breach event until mouth closure (**Figure 3**). The intrusion of marine waters into the estuary depends on numerous factors, including tide range, freshwater inflow rates, and the morphology of the channel [7–9].

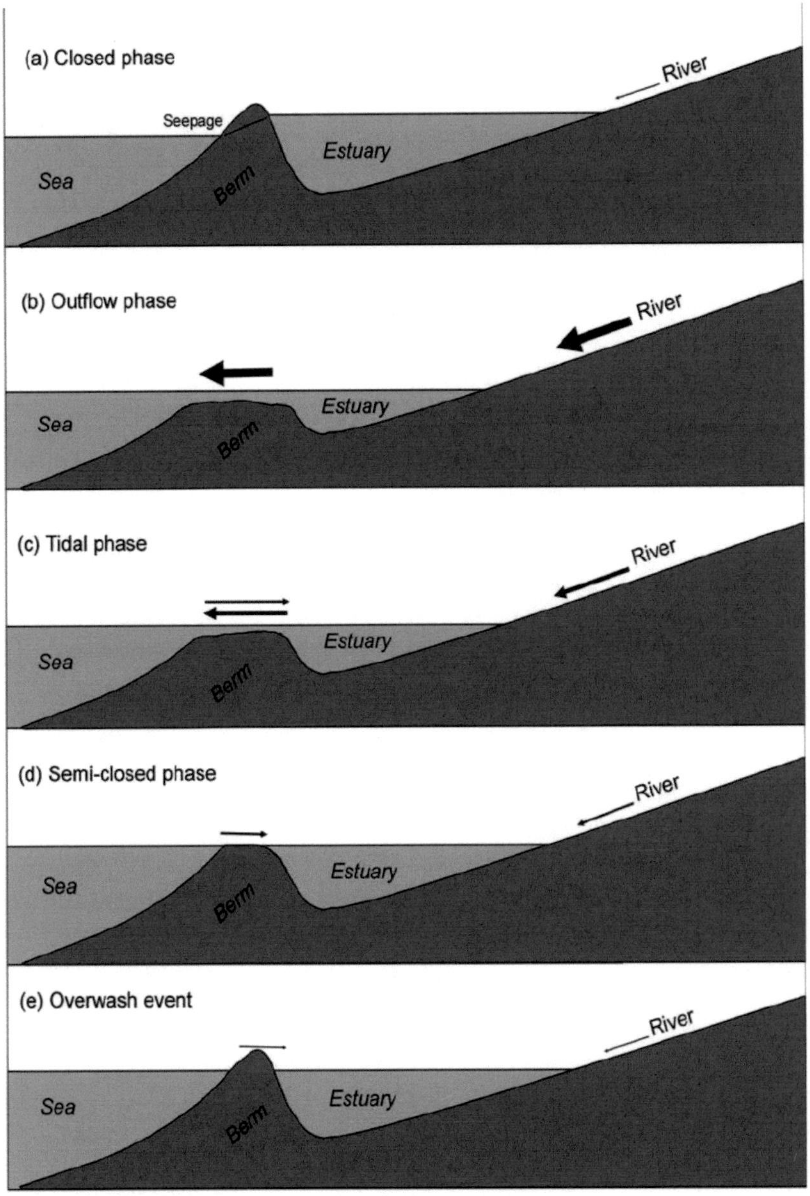

Figure 3. Hydrodynamics of perched and non-perched intermittently open/closed estuary along the southern African coastline.

The mouth dynamics of IOCEs play a key role in the overall ecosystem functioning of these systems. Intermittent breaching of the sand barriers of these systems leads to rapid changes in the physico-chemical environment, which in turn triggers major biological responses, including the out- and in-recruitment of estuarine and marine breeding invertebrates and vertebrates [2, 3]. The breaching process can also cause significant geomorphological changes because the strong breach outflows can scour large quantities of accumulated sediments from an estuary [2, 3, 9].

2. Physico-chemical environment

During closed conditions, the water temperatures in IOCEs are predominantly determined by regional climate and season and range between 18 and 30°C [7, 10–12]. Under closed conditions, salinities in IOCEs exhibit greater stability than in larger permanently open counterparts due to the lack of tidal influence [2, 12, 13]. Following rainfall, changes in salinity may be much as 30 over the course of a few days or weeks depending on freshwater input and over-washing events [12]. Mesohaline conditions (5–18) generally predominate during the closed phase, although limnetic conditions (0.1–0.5) may be recorded during periods of high rainfall [2, 3, 12, 14], while hypersaline (>40) conditions may occur during drought periods or high evaporation [1]. During the closed phase of these systems, the water column demonstrates little horizontal or vertical stratification due the reduced freshwater inflow as a result of their generally small catchment areas (<50km^2), shallow depth (generally <2 m), and strong coastal winds which facilitate the mixing of the water column [12, 14].

3. Biology

The total chlorophyll-*a* (chl-*a*) concentrations in IOCEs (0.1 and 15.4 mg chl-*a*.m^{-3}) are lower than those reported for permanently open systems (POEs) (20–100 mg chl-*a*.m^{-3}) within the same geographic region [4, 12, 15] (**Table 1**) due to reduced macronutrient availability as a result of limited freshwater inflow [12, 14, 16]. The inflow of freshwater into these systems is characterised by increased phytoplankton biomass likely sustained by the increase in macronutrient availability [12, 16]. Additionally, changes in the total chl-*a* concentration within these systems have been linked to seasonality and mouth phase [11, 14, 17]. The total phytoplankton biomass during the breaching events decreases as a result of the outflow of biologically rich estuarine waters into the marine environment [3, 14]. In contrast to the water column, microphytobenthic algae concentrations in IOCEs are two to three orders of magnitude higher than the water column phytoplankton biomass and substantially higher than those recorded in permanently open systems within the same region [4, 14, 25–28]. A combination of low turbidity, high concentrations of macronutrients in the sediments, and reduced current flow contributes to the elevated microphytobenthic algae biomass in IOCEs [4].

The total zooplankton abundance and biomass values recorded in IOCEs during the closed phase generally exceeds those levels recorded in the larger permanently open estuaries within

Estuary	Biogeographic region	Pelagic chl-a conc. ($\mu g\ L^{-1}$)	Microphytbenthic conc. (mg chl-a m^{-2})	Zooplankton abundances (ind m^{-3})
Intermittently open/closed estuaries				
Mpenjati [15]	Subtropical	0.14–15.40	19.9–616.0	8.3×10^3–8.0×10^4
Mdloti [16]	Subtropical	0.89–111.10	ND	ND
Mngazi [17]	Subtropical	ND	30–568	ND
Nyara [18]	Warm temperate	0.01–4.10	170–200	1.8×10^3–2.03×10^4
East Kleinemonde [19]	Warm temperate	0.12–6.19	ND	ND
Kasouga [12]	Warm temperate	0.29–8.01	3.87–209.9	1.06×10^3–6.1×10^4
Permanently open estuaries				
Great Fish [20]	Warm temperate	0.40–21.80	ND	ND
Kariega [21]	Warm temperate	0.64–1.13	13.6–43.8	0.9–1.6×10^3
Sundays [22]	Warm temperate	8.6–22.4	ND	1.0×10^3–5.5×10^4
Swartkops [23]	Warm temperate	4.1–8.6	ND	ND
Kromme [24]	Warm temperate	ND	ND	1.7×10^3–1.2×10^4

ND = no data

Table 1. Estimates of total water column and microphytoplankton concentrations and zooplankton abundances at selected permanently open and IOCEs within the subtropical and warm temperate biogeographic provinces along the South African coastline.

the same biogeographic region (**Table 1**) [3, 12, 29, 30]. The elevated zooplankton biomass values during closed periods are thought to be sustained by the substantial microphytobenthic stocks in these systems. Due to the limited number of marine species recorded and the virtual absence of typical estuarine zooplankton in these systems [2, 3, 31], the zooplankton communities are dominated by a few but, highly abundant hardy species, mainly copepods of the genera *Pseudodiaptomus, Acartiella,* and *Oithona* [19, 20, 32]. Changes in the zooplankton community structure within IOCEs have been linked to amongst others, mouth phase, freshwater inflow, seasonality and over-wash events [14, 32]. More recent studies indicate that predation by early life history fish may also play an important role in structuring the zooplankton communities within these systems during the close phase [33, 34]. Breaching events are typically associated with a reduction in the abundance and biomass of the zooplankton as estuarine rich waters are exported to the marine environment [14, 20]. The inflow of seawater into the estuary following the breaching event is, however, associated with an increase in the average size and zooplankton diversity as marine spawning species recruit into the system. Similarly, the over-washing of marine waters across the sandbar during spring high tides and winter storms may also contribute to an increase in the zooplankton diversity within these systems [30, 35].

Macrofaunal community composition within southern African IOCEs also demonstrates reduced diversity and is almost exclusively dominated by estuarine species [36, 37]. The low diversity can in part be attributed to the poor representation of marine breeding species to the total macrofaunal community and reduced habitat availability (submerged macrophytes and different sediment types) [36, 37]. In the virtual absence of any distinct horizontal gradients in salinity within these systems, sediment type, habitat availability, predation, and the activity of ecosystem engineers have been identified as important in determining the distribution of macrofauna within these systems [36, 37].

Due to the virtual absence of sheltered bays and increased food availability, southern African IOCEs represent important nursery areas for a large variety of marine breeding fish species [38–40]. The ichthyofaunal community structure, like the other components of the food web within these systems, is strongly determined by mouth status [38, 41–43]. Ichthyofaunal diversity and species richness during the closed phase of IOCEs are lower than those recorded in POEs within the same geographic region, due to the limited recruitment opportunities of marine breeding species associated with mouth closure and reduced habitat availability [38–40, 44]. The abundances of selected ichthofauna within IOCEs often exceed those recorded in permanently open estuaries within the same ecozone [39, 40]. The observed pattern appears largely to reflect increased food availability (mainly zooplankton and microphytoplankton stocks) within these systems. In the absence of any link to the marine environment, the ichthyofaunal community within these systems is numerically and gravimetrically dominated by estuarine species [5, 39, 44–46]. A link to the marine environment either through breaching or over-wash events has been shown to coincide with an increased contribution of marine spawning species to the total ichthyofaunal community [39, 44–46]. This has the net effect of increasing the ichthyofaunal diversity within these systems [39, 44, 45]. The magnitude of recruitment during the over-wash events is, however, substantially lower than that recorded during the open phase [3, 6, 35, 46]. The timing of the breaching events has been demonstrated to be critical in determining the magnitude of fish recruitment into these systems with maximum recruitment typically taking place in spring and summer reflecting increased availability of recruiters in the near-shore marine environment as a result of seasonal reproductive cycles [45]. The ichthyofaunal assemblages in open IOCEs are broadly similar to those recorded in the larger POEs within the same geographic region [44].

4. Anthropogenic impacts

The marine environment off the coast of South Africa is dominated by the western boundary Agulhas Current on the south-eastern seaboard and the cooler eastern boundary Benguela Current on the west coast (**Figure 1**). Like many other regions of the world, the characteristics of these currents are beginning to demonstrate the effects of global warming [47–49, 38, 39]. Sea surface temperatures (SSTs) within the Agulhas Current have increased by 1.5°C over the past four decades due to the intensification of the current caused by an increase in trade and westerly winds in the southern Indian Ocean, which have been associated with increased wind stress and curl [47]. By contrast, SSTs along the west coast have decreased resulting from the increased frequency and intensity of coastal upwelling due to upwelling favourable southerly winds [47, 49]. The warming of the Agulhas Current waters has through thermal expansion contributed to a rise in sea levels, estimated at approximately 2.74 mm y^{-1} [50], while increased wind stress has resulted in an increase in wave height (5 cm per decade), since the mid 1950s [51].

Global climate models predict that the observed changes in SST is likely to contribute to wetter conditions along the south coast of South Africa, while dryer conditions are expected along the eastern and western seaboards [52]. Moreover, global warming will be associated with an overall shortening of the wet season along the south-eastern seaboard and an increased frequency of occurrence of extreme weather events (rainfall, floods and maximum temperatures) within the region [49, 52, 53]. Regional investigations have demonstrated that the warming of marine waters along the south-eastern seaboard of southern Africa has coincided with a range expansion of warmer subtropical zone fish species into estuaries in warm temperate ecozone along the coastline [39, 49]. At the same time, there has been a concurrent decrease (10–13%) in the contribution of warm temperate species to the total fish catches within the transition zone between the two ecozones contributing to an overall decline in fish diversity within the region [39, 49]. Global climate change has probably resulted in geographic shift in the ecozone along the southern African coastline which is likely to influence connectivity and gene flow within the major phylogeographic zones between the warm temperate and subtropical ecozones [54]. It is unclear whether the observed change will be associated with a reduction in total biomass of fish within these systems, although such changes likely have knock-on effects on the importance of these systems as feeding grounds for flying birds and a source of renewable resources [55].

The impact of global climate change on the ecosystem functioning of IOCE is difficult to predict, reflecting the complex interaction between the marine and terrestrial environments. There is, however, anecdotal evidence to suggest that the impact of global climate change on the ecosystem functioning of IOCE along the South African coastline will demonstrate strong regional variability. Recent work on Intermittently Closed and Open Lakes and Lagoons (ICOLLs) along the coast of New South Wales, Australia [56] suggests that in sand deposition environments, a rise in sea level associated with climate change will increase height of the sand berm that separates the estuary from the marine environment. Such conditions are common to the warm temperate ecozone of South Africa [49]. The increase in berm height coupled with the reduction in rain season is likely to result in IOCE's being separated from the marine environment for extended periods of time. Prolonged mouth close in IOCEs is associated with a dramatic decline in species diversity and zooplankton biomass, reflecting reduced recruitment opportunities of marine breeding species into these systems (**Figure 4**) [57]. The prolonged mouth closure contributed to hypersaline conditions (salinity >40) prevailing throughout the system which was associated with decrease in the areal extent of the submerged macrophytes within the system [57]. Physiological constraints resulting from the increase temperatures and salinities and the loss of habitat (mainly submerged macrophytes) likely further contributed to the decline is species diversity observed within the systems [57, 58]. Under prolonged mouth closure, particularly during drought conditions, high evaporation rates may further increase the salinity within these systems. Under such conditions, IOCE food webs suffer a catastrophic collapse reflecting the physiological constraints of the flora and fauna within the system. Although recent models suggest that global climate change has been associated with increase in wave height, suggesting that over-wash events are likely to be more frequent; these events contribute little to the recruitment of marine breeding species into these systems [6, 35].

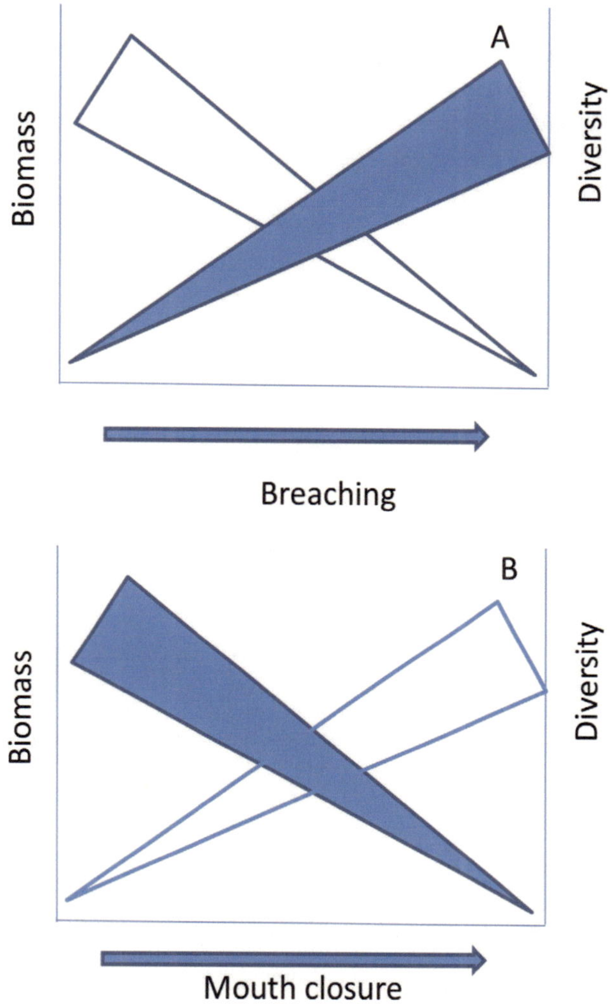

Figure 4. Predicted changes in the abundance and diversity within southern African temporarily open/closed estuaries in response to breaching (A) and prolonged mouth close (B) as a result of to global warming. Blue triangle = diversity; open triangle = biomass.

The reduced recruitment of fish into IOCE resulting from extended mouth closure may have unexpected consequences on the food web structure of these systems. Recent studies have demonstrated that the predation impact of juvenile fish, through trophic cascades, contributed to plankton food web stability within IOCE [33]. In the absence of the predators, total water column chlorophyll-a concentrations were reduced reflecting increased grazing activity by smaller heterotrophs [33, 44]. The reduced recruitment of fish into IOCEs is likely to have implications on community structure and the energy flow within these systems. Further studies are, however, required to fully appreciate the impact of changes in recruitment of the food web structure of these systems.

With regard to the IOCEs along the southern coastline of South Africa, it is likely that the projected increase in rainfall for the region coincides with the increased frequency of occurrence of breaching events, which would contribute to higher species diversity within these systems as a result of extended recruitment opportunities for marine breeding species into these systems [3]. On the other hand, frequent breaching will prevent the build-up of biomass, thus likely resulting in reduced abundances of selected components of the food web (**Figure 4**). Furthermore, the anticipated extreme flood events will likely contribute to deep mouth breaching resulting in these systems having a connection to the sea for extended periods of time.

Acknowledgements

Funds and facilities for this chapter were provided by Rhodes University.

Author details

Pierre William Froneman

Address all correspondence to: w.froneman@ru.ac.za

Department of Zoology and Entomology, Rhodes University, Grahamstown, South Africa

References

[1] Whitfield AK. A characterization of Southern African estuarine systems. Southern African Journal of Aquatic Sciences. 1992;**18**:89-103

[2] Perissinotto R, Stretch DD, Whitfield AK, Adams JB, Forbes AT, Demetriades NT. Temporarily Open/Closed Estuaries in South Africa. New York: Nova Science Publishers. 142pp. Commission Report 2010, 1247/2/04. Pretoria. 166pp

[3] Whitfield, AK, Adams JB, Bate GC, Bezuidenhout K, Bornman TG, Cowley PD, Froneman PW, Gama PT, James NC, Mackenzie B, Riddin T, Snow GC, Strydom NA, Taljaard S, Terörde AI, Theron AK, Turpie JK, van Niekerk L, Vorwerk PD, Wooldridge TH. 2008. A multidisciplinary study of a small, temporarily open/closed South African estuary, with particular emphasis on the influence of mouth state on the ecology of the system. African Journal of Marine Science. 2008;**30**(3):453-473

[4] Nozias C, Persinotto R, Mundree S. Annual cycle of microalgal biomass in a South African temporarily-open estuary: Nutrient versus light limitation. Marine Ecology Progress Series. 2001;**223**:39-48

[5] Cowley PD, Whitfield AK, Bell KNI. The surf zone ichthyoplankton adjacent to an intermittently open estuary, with evidence of recruitment during marine overwash events. Estuarine, Coastal and Shelf Science. 2001;**52**:339-348

[6] Kemp JOG, Froneman PW. Recruitment of ichthyoplankton and macrozooplankton during overtopping events into a temporarily open/closed southern African estuary. Estuarine, Coastal and Shelf Science. 2004;**61**:529-537

[7] Harrison TD. Ichthyofauna of South African estuaries in relation to the zoogeography of the region. Smithiana Bulletin. 2005;**6**:1-27

[8] Cooper JAG. Geomorphological variability among microtidal estuaries from the wave dominated South African coast. Geomorphology. 2001;**40**:99

[9] Whitfield AK, Bate GC, Adams JB, Cowley PD, Froneman PW, Gama PT, Strydom NA, Taljaard S, Theron AK, Turpie JK, van Niekerk L, Wooldridge TH. A review of the ecology and management of temporarily open/closed estuaries in South Africa, with particular emphasis on river flow and mouth state as primary drivers of these systems. African Journal of Marine Science. 2012;**34**(2):163-180

[10] Day JH. Estuarine Ecology with Particular Reference to Southern Africa. Cape Town: A.A. Balkema; 1981. 411pp

[11] Perissinotto R, Blair A, Connell A, Demetriades NT, Forbes AT, Harrisson TD, Iyer K, Joubert M, Kibirige I, Mundree S, Simpson H, Stretch D, Thomas C, Thwala X, Zietsman I. Contributions to Information Requirements for the Implementation of Resource Directed Measures for Estuaries (Volume 2). Responses of the biological communities to flow variation and mouth state in two KwaZulu-Natal temporarily open/closed estuaries. Water Research Commission 2004

[12] Froneman PW. Seasonal changes in selected physico-chemical and biological variables in the temporarily open/closed Kasouga estuary, Eastern Cape, South Africa. African Journal of Aquatic Science. 2002;**27**:117-123

[13] Begg GW. The comparative ecology of Natal's smaller estuaries. Natal Town and Regional Planning Report. 1984;**62**:182

[14] Froneman PW. Response of the biology to three different hydrological phases in the temporarily open/closed Kasouga estuary. Estuarine, Coastal and Shelf Science. 2002; **55**:535-546

[15] Kibirige I, Perissinotto R, Thwala X. A comparative study of zooplankton dynamics in two subtropical temporarily open/closed estuaries, South Africa. Marine Biology. 2006;**148**: 1307-1324

[16] Thomas CM, Perissinotto R, Kibirige I. Phytoplankton biomass and size structure in two South African eutrophic, temporarily open/closed estuaries. Estuarine coastal and Shelf Science. 2005;**65**:223-238

[17] Mbande S, Froneman PW, Whitfield AK. A preliminary assessment of the primary carbon sources utilised by fishes in the Mnagazi and Mngazana estuaries. African Journal of Aquatic Sciences. 2004;**29**:195-204

[18] Perissinotto R, Walker DR, Webb P, Wooldridge TH, Bally R. Relationships between zoo- and phytoplankton in a warm-temperate, semi-permanently closed estuary, South Africa. Estuarine Coastal and Shelf Science. 2000;**51**:1-11

[19] Froneman PW. Seasonal patterns in the zooplankton community structure and biomass in an Eastern Cape temporarily open/closed estuary. Estuarine Coastal and Shelf Science. 2004;**60**:125-132

[20] Grange N, Whitfield AK, De Villiers CJ, Allanson BR. The responses of two estuaries on the Eastern Cape coast of South Africa to varying freshwater inflow: Implications for environmental management of freshwater resources. Aquatic Conservation: Marine and Freshwater Ecosystems. 2000

[21] Richoux NB, Froneman PW. Assessment of spatial variation in carbon utilisation by benthic and pelagic invertebrates in a temperate South African estuary using stable isotope signatures. Estuarine, Coastal and Shelf Science. 2007;**71**:545-558

[22] Jerling HL, Wooldridge TH. Plankton distribution and abundances in the Sundays river estuary, South Africa with comments on potential feeding interactions. South African Journal of Marine Science. 2010;**15**:169-185

[23] Scharler UM, Baird D. The influence of catchment management on salinity, nutrient stochiometry and phytoplankton biomass of Eastern Cape estuaries, South Africa. Estuarine Coastal and Shelf Science. 2003;**56**:735-748

[24] Wooldridge TH, Callahan R. The effects of a single freshwater release into the Kromme Estuary. 3: Estuarine zooplankton response. Water SA. 2000;**26**:311-318

[25] Adams JB, Bate GC. Primary producers: Estuarine microalgae. In: Allanson BR, Baird D, editors. Estuaries of South Africa. Cambridge: University Press; 1999. pp. 91-100

[26] Adams JB, Bate G, O'Çallaghan MO. Primary producers. Estuarine macrophytes. In: Allanson BR, Baird D, editors. Estuaries of South Africa. Cambridge: University Press; 1999. pp. 100-117

[27] Campbell EE, Knoop WT, Bate CG. A comparison of phytoplankton biomass and primary production in three Eastern Cape estuaries. South African Journal of Science. 1991; **87**:259-264

[28] Skinner T, Adams JB, Gama PT. The effect of mouth opening on the biomass and community structure of microphytobenthos in a small oligotrophic estuary. Estuarine, Coastal and Shelf Science. 2006;**70**:161-168

[29] Kibrige I, Perissinotto R. The zooplankton community of the Mpenjati Estuary, a South African temporarily open/closed system. Estuarine, Coastal and Shelf Science. 2003;**58**: 727-741

[30] James NC, Cowley PD, Whitfield AK. Abundance, recruitment and residency of two sparids in an intermittently open estuary in South Africa. African Journal of Marine Science. 2007;**29**(3):527-538

[31] Froneman PW. Food web structure in three contrasting estuaries determined by stable carbon isotope analysis. African Journal of Aquatic Sciences. 2002;**27**:107-115

[32] Grindley JR. Estuarine plankton. In: Day JH, editor. Estuarine Ecology, with Particular Reference to Southern Africa. Cape Town: A.A. Balkema; 1981. pp. 117-146

[33] Wasserman RJ, Noyon M, Avery TS, Froneman PW. Trophic level stability-inducing effects of predaceous early juvenile fish in an estuarine mesocosm study. PLoS One. 2013; **8**:e61019

[34] Wasserman RJ, Vink TJF, Kramer R, Froneman PW. Hyperbenthic and pelagic predators regulate alternate key planktonic predators in shallow temperate estuaries. Marine and Freshwater Research. 2014;**65**:791-801

[35] Bernard ATF, Froneman PW. Population dynamics of the caridean shrimp *Palaemon peringueyi* in a temperate temporarily open/closed southern African estuary. Estuarine, Coastal and Shelf Science. 2005;**63**:73-81

[36] Teske PR, Froneman PW, Barker NP, McQuaid CD. Phylogeographic structure of the caridean shrimp, *Palaemon peringueyi,* in South Africa: Further evidence for intraspecific genetic units associated with marine biogeographic provinces. African Journal of Marine Science. 2007;**29**(2):253-258

[37] Teske PR, McQuaid CD, Froneman P, Barker NP. Impacts of marine biogeographic boundaries on phylogeographic patterns of three South African estuarine crustacean species. Marine Ecology Progress Series. 2006;**314**:283-293

[38] Whitfield AK. Biology and ecology of fishes in southern African estuaries. Ichthyological Monographs of the J.L.B Smith Institute of Ichthyology. 1998;**2**:223

[39] James NC, Cowley PD, Whitfield AK, Lamberth SJ. Fish communities in temporarily open/closed estuaries from the warm- and cool-temperate regions of South Africa: A review. Reviews in Fish Biology and Fisheries. 2007;**17**:565-580

[40] James NC. Trends in fish community structure and recruitment in a temporarily open/ closed South African estuary [PhD Thesis]. Grahamstown: Rhodes University; 2007. 171pp

[41] Bennett BA, Hamman KCD, Branch GM, Thorne SC. Changes in the fish fauna of the Bot River Estuary in relation to opening and closure of the estuary mouth. Transactions of the Royal Society of South Africa. 1985;**45**:449-464

[42] Harrison TD, Whitfield AK. Fish community in three temporarily open/closed estuaries on the Natal coast. Ichthyological Bulletin. 1995;**64**:1-80

[43] Lukey JR, Booth AJ, Froneman PW. Fish population size and movement patterns in a small intermittently open South African estuary. Estuarine, Coastal and Shelf Science. 2006;**67**:10-20

[44] Vorwerk PD, Whitfield AK, Cowley PD, Paterson AW. The influence of selected environmental variables on fish assemblage structure in a range of southeast African Estuaries. Environmental Biology of Fishes. 2003;**66**(3):237-247

[45] Bell KNI, Cowley PD, Whitfield AK. Seasonality in frequency of marine access to an intermittently open estuary: Implications for recruitment strategies. Estuarine, Coastal and Shelf Science. 2001;**52**:327-337

[46] Tweddle GP, Froneman PW. Influence of mouth status on population structure of southern African endemic estuarine-spawning ichthyofauna in a temperate, temporarily open/ closed estuary. African Journal of Aquatic Science. 2015;**40**(2):221-225

[47] Rouault M, Penven P, Pohl B. Warming in the Agulhas Current system since the 1980's. Geophysical Research Letters. 2009;**36**:L12602

[48] Moloney CL, Fennessy ST, Gibbons MJ, Roychoudhury A, Shillington FA, von der Heyden S, Watermeyer K. Reviewing evidence of marine ecosystem change off South Africa. African Journal of Marine Science. 2013;**35**:427-448

[49] Mead A, Griffiths CL. et al. Human-mediated drivers of change-impacts on coastal ecosystems and marine biota of South Africa. African Journal of Marine Science. 2013;**35**:403-425

[50] Mather A, Garland G, Stretch D. 2009 – Southern African sea levels: Corrections influences and trends. African Journal of Marine Science. 2009

[51] Trenberth KE, Jones PD, Ambenje P, Bojariu R, Easterling D, Klein-Tank A, Parker D, Rahimzadeh D, Renwick JA, Rusticucci M, Sonden B, Zhai P. Observations: Surface and atmospheric climate change In: Solomon S, Qin D, Manning M, Chen Z, Marquis M, Averyt KB, Tignor M, Miller HL, editors. Climate Change 2007: The Physical Science Basis. Contribution of working group 1 to the fourth assessment report of the Intergovernmental Panel on Climate Change. United Kingdom: Cambridge University Press; 2007

[52] Hewitson BC, Tennant W, Walawege R. Atmospheric moisture transport and sources for southern Africa. Water Research Commission Report 1012/1/04. Pretoria. 2004. 88pp

[53] New M, Hewitson B, Stephenson DB, Tsisga A, Kruger A, Manhique A, Gomez B, Coelho CAS, Masisi DN, Kululanga E, Mbambalala E, Adesina F, Saleh H, Kanyanga J, Adosi J, Bulane L, Fortunanta L, Mdoka ML, Lajoie R. Evidence of trends in daily climate extremes over Southern and West Africa. Journal of Geophysical Research. 2006;**111**:1-11

[54] James NC, Whitfield AK, Cowley PD. Preliminary indications of a climate induced change in a warm temperate South African estuarine fish community. Journal of Fish Biology. 2008;**72**:1855-1863

[55] Terorde AL, Turpie JK. Use of a small intermittently-open estuary by water birds: A case study of the East Kleinemonde Estuary, Eastern Cape, South Africa. African Journal of Aquatic Science. 2012;**37**:183-190

[56] Haines PE, Thom BG. Climate change impacts on entrance processes of intermittently open/closed coastal lagoons in new South Wales, Australia. Journal of Coastal Research. 2007;**50**:242-246

[57] Froneman PW, Henninger T. The influence of prolonged mouth close on selected components of the hyperbenthos is the temporarily open/closed Kasouga Estuary, South Africa. Estuarine, Coastal and Shelf Science. 2009;**83**:326-332

[58] Henninger T, Froneman PW, Richoux N, Hodgson AN. Role of submerged macrohytes and a refuge and food source for the estuarine isopod, *Exospheroma hylocoetes*. Estuarine, Coastal and Shelf Science. 2009;**82**:285-293

Chapter 6

The Relationship of Sediment and Intersitial Water Properties with Mangrove Health in a Subtropical Coastal Lagoon of Mexico

Carlos Augusto Zenteno-Palma, Aramis Olivos-Ortiz,
María del Carmen Álvarez,
Sonia Isabel Quijano-Scheggia and
Gloria Alicia Jiménez-Ramón

Additional information is available at the end of the chapter

http://dx.doi.org/10.5772/intechopen.73045

Abstract

The aim of this study was to determine the influence of the physicochemical properties of interstitial water, sediment geochemistry and sediment grain size on the health of mangroves in three Basins (I, II and III) in Cuyutlán Lagoon, Mexico. Monthly sampling was conducted at eight stations from February to October 2014. The optimum reforestation conditions were observed at stations in Basin I (1A and 1B) and II (2C and 2D), where exchange of water between the ocean and lagoon predominates (Tepalcates and Ventanas Channels) and tidal influence induces water exchange that generates increased dissolved oxygen (3–4 mg/l) concentrations, low salinity (20–30) and greater coverage of healthy mangrove, with sufficient organic carbon (4–6%) for assimilation by the plants. By contrast, stations in Basin III were characterised by water stagnation resulting in increased salinity (40–50), depletion of the dissolved oxygen concentrations (0.1–2 mg/l) and low organic carbon (2%) in sediments, which contributed to low dwarf mangroves with low coverage most likely in response to environmental stress. Finally, the healthiest mangrove was found in areas where the dynamics favoured the deposition of organic carbon and medium sands in the sediment, which generated greater nutrient availability for fixation and assimilation by the roots.

Keywords: Cuyutlán Lagoon, physicochemical, organic carbon, interstitial water, sediment

1. Introduction

Coastal lagoons are considered geomorphological depressions in the coastal area less than 10 m deep, protected by physical barriers or hydrodynamic processes [1]. These coastal

INTECH
open science | open minds

© 2018 The Author(s). Licensee InTech. This chapter is distributed under the terms of the Creative Commons Attribution License (http://creativecommons.org/licenses/by/3.0), which permits unrestricted use, distribution, and reproduction in any medium, provided the original work is properly cited.

environments generally have permanent or ephemeral links with the sea and have unique physical, chemical and biological conditions resulting from the interactions of two different bodies of water [2, 3]. These ecosystems are of great economic and ecological importance due to high primary productivity and by providing areas for breeding, feeding and protection of diverse species [4].

The elevated productivity in these water bodies is related to the input of inorganic nutrients resulting from ocean-lagoon and continent-lagoon interactions, and the recycling of labile organic matter (OM) deposited in the sediments which plays a fundamental role in the formation of assailable nutrient product diagenetic processes (remineralisation) of organic matter from primary producers [5, 6]. Moreover, the particle size distribution and the proportion of organic content affect the distribution of organisms in sediments and the zonation of aquatic vegetation. In addition, based on its structure and composition, the OM accumulated in sediments can provide a record of primary productivity fluctuations in these marine environments [7, 8].

Sedimentation rates in coastal lagoons vary widely, and the type of sediments associated with these systems is considered an important ecological factor, due to concentrations of trace elements, nutrients and OM necessary for basic physiological processes of primary producers [9]. Furthermore, biogeochemical cycles in these systems are affected by geochemical processes between the sediment-water interface such as exchange and diffusion of dissolved ions and gases [10].

Mangrove ecosystems are characterised as wetlands located in tropical and subtropical coastal areas of the world and are associated with complex sedimentary environments such as estuaries and coastal lagoons. These systems are major producers of OM and serve as ecologically important protected areas and breeding and nursery grounds for many species [11–15].

The presence of mangrove communities in the coastal zone provides important ecosystem services including protection from storms surges, protection against coastal erosion by wave activity and provision of biological filters which retain and process contaminants and nutrient excess, organic matter decomposition. In addition, they act as sinks for CO_2, potentially mitigating climate change [16]. This type of wetland is characterised by halophile properties [17] that withstand wide variations in salinity and temperature in short periods of time [15]. Mangrove forests are associated with conditions such as silt-clay sediments with high accumulation of organic matter [18], sediment flooding for better propagule rooting [19] and high, constant nutrient availability [20–23].

Because of the important ecological and environmental services provided by mangroves, mitigation measures are necessary to avoid deforestation of these ecosystems, as well as restoration activities (replanting). For successful mangrove restoration, knowledge of the surface and subsurface physicochemical conditions of the sediments is extremely important. The objective of this study is to generate a baseline of physicochemical and sediment data, which will improve mangrove replanting activities in this tropical lagoon and serve as a tool to promote restoration in other similar regions.

2. Materials and methods

2.1. Description of the study area

Cuyutlán Lagoon (19°07′ N, 104°05′ W) is located on the northeast coast of Colima, Mexico, bordered to the north by Manzanillo Bay and by Armeria River to the south. Through natural physiography and engineering activities, the lagoon system was divided into three Basins (I, II and III, **Figure 1**), with a large expanse of mangrove forest the edge of each basin, primarily consisting of *Laguncularia racemosa* and *Rhizophora mangle*. The region covers an area of approximately 7200 hectares, with a length of 37 km and a 5–6 km wide sand bar [22]. The lagoon is connected to the sea by three channels: Tepalcates, Ventanas and Tunnel [24]. The study area has a warm, humid tropical climate Aw_0 type, with an annual temperature variation between 25 and 26°C. The summer rainy season begins around May and ends in October, with the highest rainfall occurring in July and September; winter has a lower rainfall of 5 mm, with an average annual rainfall between 800 and 1200 mm [25].

2.2. Field methodology

From February to October 2014, monthly sampling was carried out in three stations occupied in three basins of the Cuyutlán Lagoon, which were established based on areas of mangrove replanting. Two stations were located in Basins I and II (1A and 1B and 2C and 2D, respectively), while four stations were located in Basin III (3E, 3F, 3G and 3H). Samples were taken

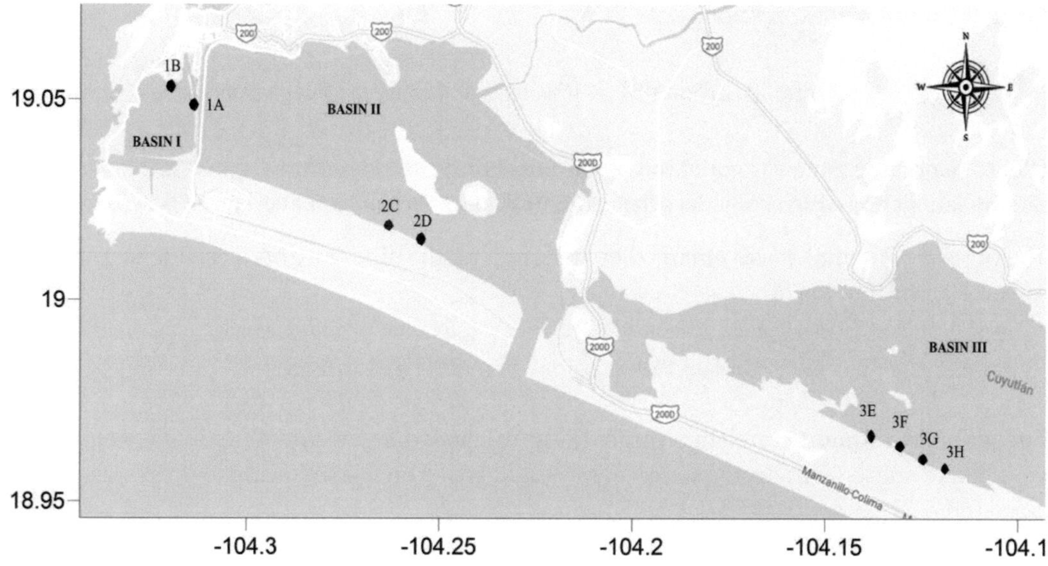

Figure 1. Geographical location and reforestation areas (Basins I, II and III) analysed in Cuyutlán Lagoon, Colima, Mexico.

to determine physicochemical parameters of the sediment pore water reforested areas, during flooded and nonflooded conditions. In the nonflooded stations, a pit of 15–20 cm depth was excavated which was allowed to fill with interstitial water. For flooded stations, the measurements were made in the water adjacent to the sediment. Temperature, dissolved oxygen and salinity were measured using a mini CTD YSI Pro 2030, while the pH was measured with an Orion Star Thermo Scientific potentiometer A221.

Sediment cores were extracted manually with PVC tubes 1 m length and 4.5 cm in diameter for textural analysis (grain size, porosity and humidity) and 3 cm in diameter for chemical analysis (percent organic carbon) during February, March, May, June and August.

To establish the current health status of the mangroves, they were evaluated according to the methodology proposed by [26]. Fifteen indicators were used: leaf colouration, leaf cover, plant height, phenological aspects (flowering, fruiting, or vegetative aspect in later stages) time to reproduction (if not mature), number of propagules or reproductive structures (e.g. few flowers or a few propagules on the ground), presence of dwarfism, turgor and brilliance of leaves, stem strength, presence of pest damage, presence of salty plains, presence of marshes, diameter at breast height (DBH) of less than 10 cm and a-zonal distribution.

2.3. Laboratory methodology

The organic carbon content of the sediment was determined by the 2540G method proposed by [27], a semi-quantitative determination which consists of drying the sediment at 105°C for 3 hours, after which the dry sample is incinerated at 550°C in a Felisa FE340 muffle. The percentage organic carbon in the sediment was obtained by the weight difference in the sample using the following expression:

$$\text{Organic carbon (\%)} = \left[(\text{\%Organic Matter})/1.724\right] * 100 \tag{1}$$

Where % organic matter is equal to volatile substances in the sediments and 1.724 is the Van Bemmelen factor, which considers that organic matter contains on average 58% carbon.

The porosity of samples was obtained by determining the difference between the wet and dry samples according to [28].

$$\text{Porosity (\%)} = \text{Initial weight} - \text{Final weight} \tag{2}$$

The sediment humidity (moisture content) was calculated by the difference in the wet sample weight and subsequent drying to constant weight based on the following expression:

$$\text{Humidity(\%)} = \left[(\text{Initial weight} - \text{Final weight})/(\text{Initial weight})\right] * 100 \tag{3}$$

A descriptive analysis of the data (mean, variance and standard deviation) was performed followed by tests of normality (Kolmogorov-Smirnov and Lilliefors test) and homoscedasticity of variance (Cochran, Hartley and Bartlett test) to define whether to use parametric or nonparametric analyses. To establish the relationship of elemental chemical content with sediment texture, a simple correlation was performed at a significance level of $p < 0.05$, in addition to a descriptive data analysis of normality and homoscedasticity to establish possible differences between sampling stations.

3. Results

The mean concentration of interstitial salinity in mangrove reforestation areas varied between 25 and 30 during the study (**Figure 2**). The highest salinity values were recorded in Basin III (3E, 3F, 3G and 3H) stations, with minimum of 5 and maximum of 45–50. On the other hand, stations in Basin I (1A y 1B) recorded minimum 12 and maximum close to 40, and the same pattern was observed in Basin II (2C y 2D) stations with minimum values of 10 and maximum of 35. Spatially, the widest range of salinity was detected in Basin III, while Basin II showed the least variation in values. Temporally (**Figure 2B**), the highest salinity concentrations were recorded during the months of April (41), May (39) and June (34), gradually decreasing until September (10); months were significantly different ($p < 0.05$). In the dry season, salinity reached the highest average (35), with a range from 10 to 50; on the other hand, during the rainy season, the average salinity was 24 (range 5–38). The seasons were significantly different ($p = 0.00$).

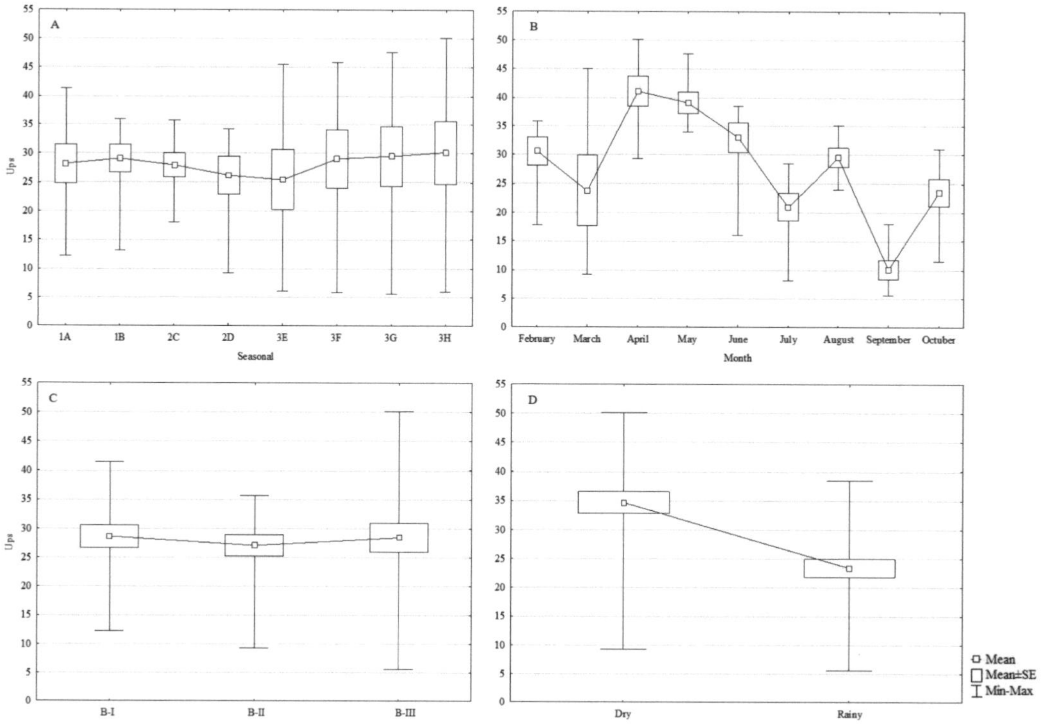

Figure 2. Seasonal (A), monthly (B), spatial (C) and temporal (D) variation of the salinity in interstitial water on reforestation zones of Cuyutlán Lagoon.

The average temperature obtained in the areas of reforestation was 29–30°C. The largest temperature fluctuations occurred in station 1A (Basin I) with min of 25°C and max of 35°C. At the remaining stations, temperatures did not exceed 32°C. There were no significant spatial differences in temperature evident in the replanted areas ($p > 0.05$) (**Figure 3A**). Monthly variations in temperature demonstrated significant differences ($p < 0.05$), with the highest temperature fluctuation in March from 25 to 35°C. A significant increase was recorded in May, which oscillated between 28 and 31°C until October. Low temperatures were obtained in February and April, with significant differences ($p < 0.05$) (**Figure 3B**). Spatially, Basin I showed the greatest variability in temperature, with a maximum of 35°C, while the minimum of 24.3°C was recorded in the Basin III (**Figure 3C**). Significant differences in temperature ($p < 0.05$) were recorded between the rainy season (mean of 30°C) and dry season (mean of 24°C) (**Figure 3D**).

The mean values of interstitial dissolved oxygen concentrations registered in the stations in Basins I and II ranged between 2 and 4 mg/l (**Figure 4A**), while in Basin III, the average values ranged between 0.1 and 2 mg/l. The greatest fluctuations were recorded at stations 1A and 1B, both with a minimum of 0.1 mg/l and maximum values of 8 and 10 mg/l, respectively. The maximum values in Basin III did not exceed 2 mg/l. Dissolved oxygen concentration in Basin III was statistically different from those recorded Basins I and II ($p > 0.05$). Monthly records showed an increase in dissolved oxygen concentrations in May (4 mg/l), a significant variation compared to previous months (February, March and April), which had lower values (range 0–0.1 mg/l).

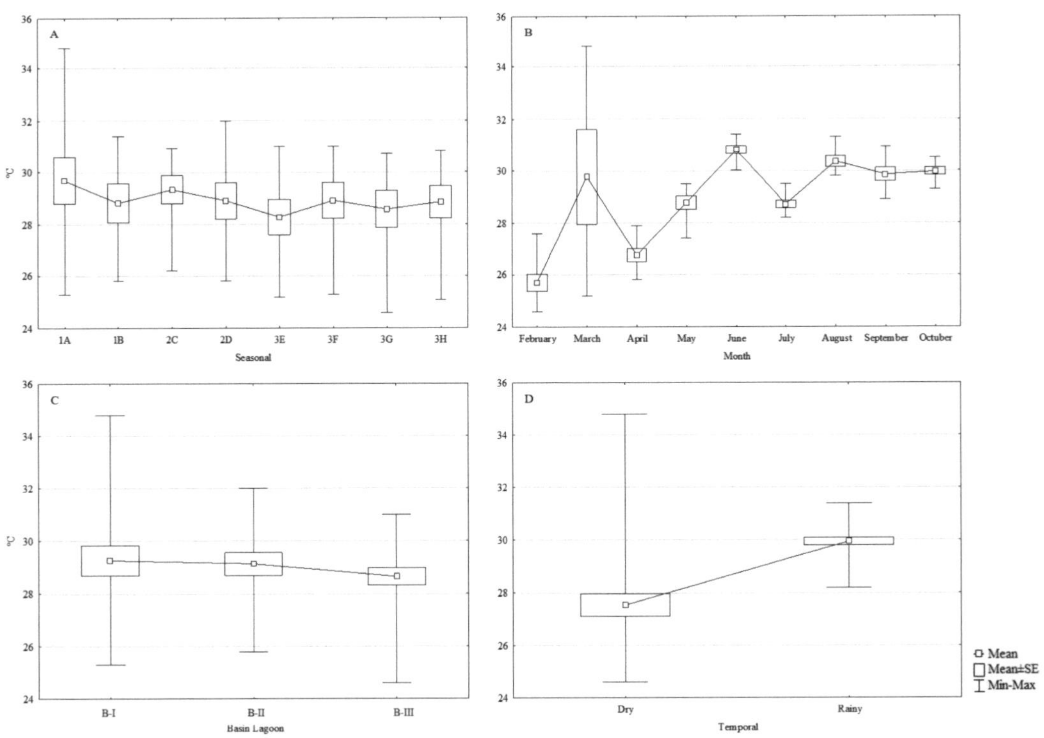

Figure 3. Seasonal (A), monthly (B), spatial (C) and temporal (D) variation of the temperature in interstitial water on reforestation zones of Cuyutlán Lagoon.

September showed the highest fluctuations, with a minimum of 0.1 mg/l and maximum of 10 mg/l (**Figure 4B**). Statistically, there were no significant differences in dissolved oxygen concentrations between February, March and April ($p > 0.05$). Basin III presented low oxygen concentration with an average of 0.2 mg/l, whereas Basins I and II maintained an average between 2 and 4 mg/l (**Figure 4C**). No significant seasonal differences occurred; however, the oscillations were highest during the rainy season (min = 0.1 mg/l; max = 10 mg/l) (**Figure 4D**).

The pH interstitial values showed no statistically significant variation between stations ($p > 0.05$). On average, pH values ranged from 7.8 to 8.1 and demonstrated no significant spatial patterns ($p > 0.05$). Similarly, there were no significant differences in the pH values between the months of February to April ($p > 0.05$). Station 1A registered a pH minimum of 7 and maximum of 8.6, whereas in station 1B, the minimum was 6.9 and maximum 9.1 (**Figure 5A**). Monthly values were homogenous between February and April, averaging 7.7. On the other hand, pH increased significantly to 8.3 in May ($p < 0.05$). Values decreased from May to July, until reaching the minimum value recorded in the study (7.2). Thereafter, the values progressively increased until October (8.4) (**Figure 5B**). Spatially, the different basins (**Figure 5C**) did not register a significant difference ($p > 0.05$); nevertheless, Basin I showed greater pH variation, with minimum values of 6.9 and maximums of 9.1. In Basins II and III, pH values on average ranged between 7.0 and 8.4. Finally, there were no significant differences between seasons; nevertheless, greater pH variability was observed in the rainy months (**Figure 5D**).

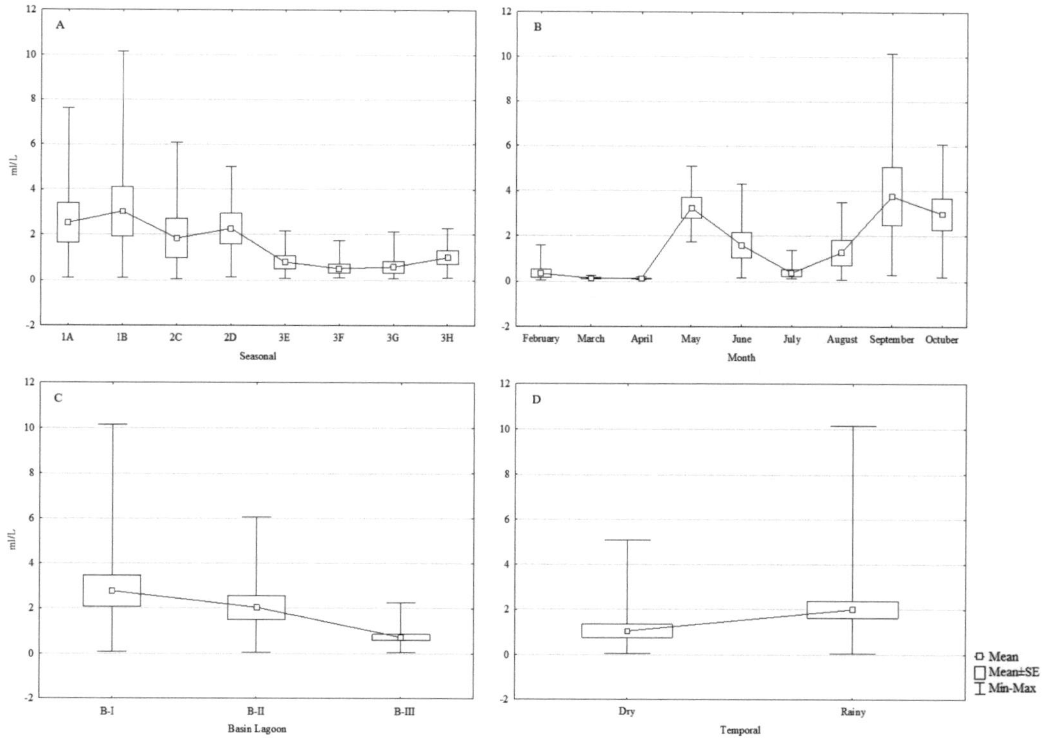

Figure 4. Seasonal (A), monthly (B), spatial (C) and temporal (D) variation of the dissolved oxygen in interstitial water on reforestation zones of Cuyutlán Lagoon.

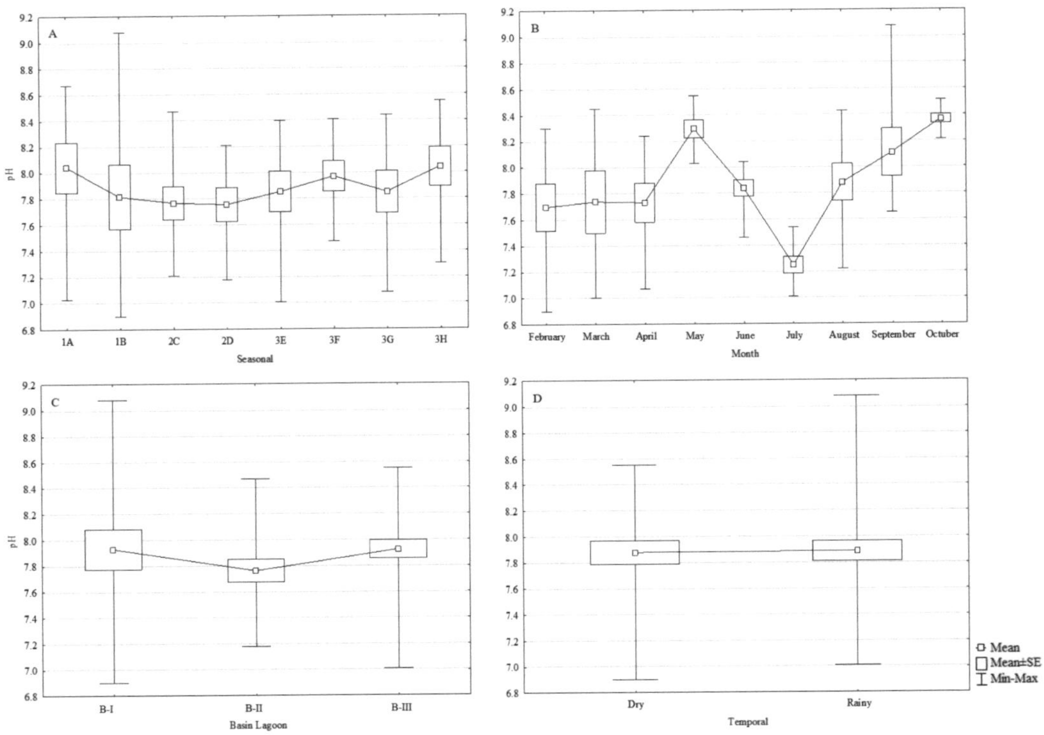

Figure 5. Seasonal (A), monthly (B), spatial (C) and temporal (D) variation of pH in interstitial water on reforestation zones of Cuyutlán Lagoon.

With regard to the percentage organic carbon (OC), the highest values during the sampling period were recorded stations 1A, 1B, 2C, 2D and 3E. On the other hand, the months with the highest OC content on average were March (5.9%), May (5.6%) and August (5.5%) (**Figure 6B**). In February, the maximum value was obtained in station 1B (7.7%) and the minimum at station 3H (1.6%) (**Table 1, Figure 6A**). In March and May, the maximum concentrations were observed in stations 1A, 1B and 2D with average values of 6.5%, while the minimum values were recorded in station 2C with 3% (**Figure 6A**). In June and August, high concentrations were recorded in Basins I and II (7.7% and 5.9%), while in Basin III, percentages were relatively low (2.7% and 3.2%). In general, Basin I showed higher organic carbon content (7.2%) in the reforestation zones compared to Basin III (3.8%) (**Figure 6C**). There were no significant temporal differences in the percentage organic carbon in the sediments evident during the investigation ($p > 0.05$) (**Figure 6D**).

Table 1 shows the mean values obtained from the textural analysis (grain size, porosity and humidity) and the organic carbon content of the sediment. The predominant grain size during the study showed no significant spatial variations and was comprised of medium sands (MS). The sediment porosity values averaged between 1.2 and 1.4 at stations in Basin I (1A and 1B). At those stations in Basins II (2C and 2D) and III (3E, 3F, 3G and 3H), the mean porosity values were lower and ranged between 0.4 and 0.8. Humidity showed a significant variation between the stations ($p < 0.05$). During February, the humidity was between 58.9 and 65.8% (**Table 1**), which decreased to 30.2% in March in Basin I, while in Basin II, higher percentages were preserved (78.1%). In May, this pattern continued with low humidity values in Basin I and high values in Basin II. During June, the humidity values showed the same behaviour obtained in February,

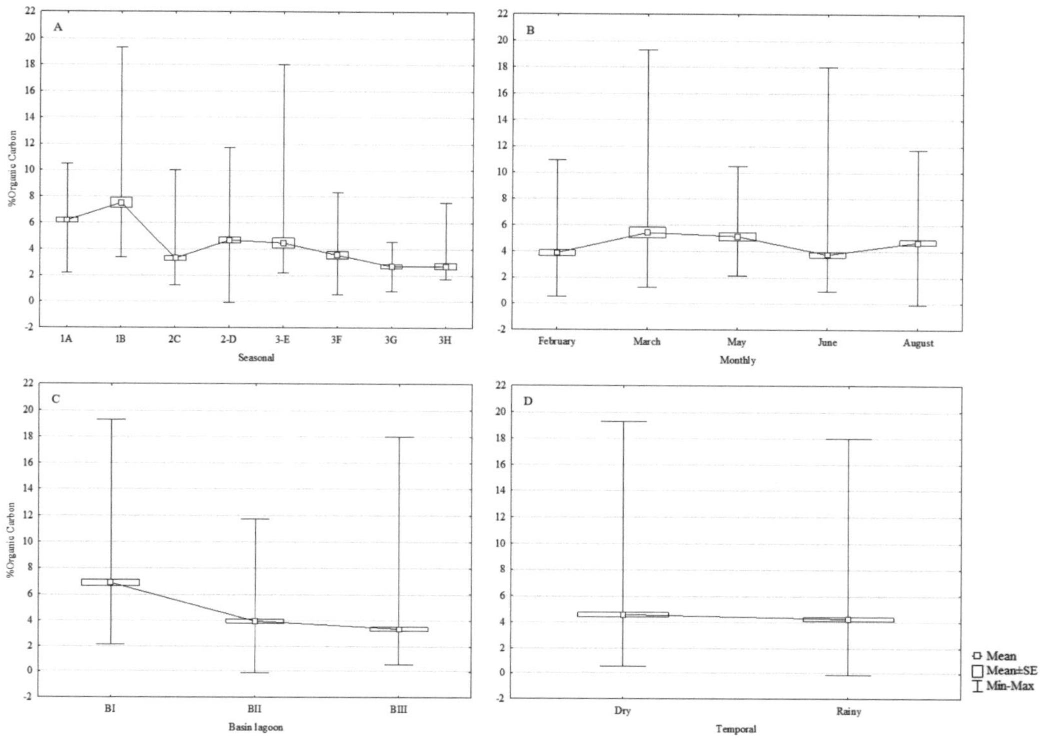

Figure 6. Seasonal (A), monthly (B), spatial (C) and temporal (D) variation of organic carbon content in sediments of Cuyutlán Lagoon.

with high values in all the stations. The maximum percentage obtained was found in station 3H with 71.6% and the minimum in station 2D with 62.5% (**Table 1**).

In the replanting zones in the Cuyutlán Lagoon, two species of mangrove were recorded in Basins I and II, *L. racemosa* and *R. mangle*, while in Basin III, only *L. racemosa* was recorded. In 2012, mangrove health was classified into three types: weak, vigorous and very vigorous. In Basin I, no results were obtained due to the lack of mangroves in the study areas. However, in Basin I, the mangrove was classified as vigorous in the southern area, while the northern area growth was deemed to be weak. Basin III mangroves were in a vigorous to very vigorous state in the areas near the shore (both in the north and south of the lagoon).

Throughout the investigation (2015), an evolution of the mangrove ecosystem was observed in each of the replanting zones of the Cuyutlán Lagoon. At those stations in Basin I (1A and 1B), the mangrove was classified as vigorous with an initial reproduction process, so that the height of the forest in these stations was between 2 and 2.5 m, maintaining a very full and extensive coverage. Similarly, at Basin II (stations 2C and 2D), a very vigorous mangrove forest was also observed with mangrove heights of between 2.5 and 4 m which were in an advanced reproduction state. It is worth mentioning that at these stations, the greatest coverage and suitable mangrove growth were observed. By contrast, the mangroves in Basin III, particularly at stations 3F, 3G and 3H, were characterised as shrub growth and excessive salt secretion. To the contrary, station 3E presented optimal growth with the mangrove classified as vigorous.

Date	Parameter	Station							
		1A	1B	2C	2D	3E	3F	3G	3H
February	% organic carbon	6.940	7.692	3.365	4.423	4.036	2.727	1.653	1.598
	Textural group	MS	MS	MS	MS	MS	MS	MS	MS
	Porosity (%)	1.281	1.2880	0.702	0.771	0.844	0.715	0.583	
	Humidity (%)	62.61	62.74	65.86	62.47	58.91	65.17	71.63	
March	% organic carbon	6.057	8.902	2.050	5.669	Dry	Dry	Dry	Dry
	Textural group	MS	MS	MS	MS	–	–	–	–
	Porosity (%)	1.433	1.327	0.447	0.958	–	–	–	–
	Humidity (%)	30.39	35.26	78.12	55.53	–	–	–	–
May	% organic carbon	7.594	6.962	4.127	3.252	Dry	Dry	Dry	Dry
	Textural group	MS	MS	MS	MS	–	–	–	–
	Porosity (%)	1.409	1.381	0.826	0.891	–	–	–	–
	Humidity (%)	34.1	34.98	58.89	57.76	–	–	–	–
June	% organic carbon	5.053	5.906	4.210	3.177	4.114	2.846	3.247	2.718
	Textural group	MS	MS	MS	MS	MS	MS	MS	MS
	Porosity (%)	1.463	1.457	0.827	0.937	0.733	0.763	0.723	0.623
	Humidity (%)	66.56	65.47	62.5	57.61	66.27	64.94	67.17	71.62
August	% organic carbon	5.858	7.692	2.941	5.608	5.212	5.113	3.247	2.718
	Textural group	MS	MS	MS	MS	MS	MS	MS	MS
	Porosity (%)	1.381	1.317	0.534	0.858	0.458	0.458	0.723	0.623
	Humidity (%)	32.91	35.75	73.89	60.16	79.08	85.92	67.17	71.62

Table 1. Monthly results of organic carbon, textural group, porosity and humidity of the sediments in the different mangrove replanting areas of Cuyutlán Lagoon (MS = Medium sand).

4. Discussion

The reforestation zones showed a marked seasonal variation in the physicochemical variables during the study. Salinity in Basins I (1A and 1B), II (2D, 2C) and III (3E, 3F, 3G, 3H) were highest in summer (April-August), before the onset of the rainy season in September and October contributed to decreased values [29]. During the summer months, the volume of water inside the lagoon was reduced, and no sediment rinsing occurred causing an excessive accumulation of salts on the surface and generating dehydration in the plants, which is associated with biological stress. It has been reported that some mangrove species are not tolerant to elevated salinity or to desiccation conditions, which causes them to weaken and reduce competition against other vascular plant species [30]. Also, due to the high evaporation rates in this period, there is an increased availability of dissolved metallic cations, which are absorbed from the sediment by the roots of the mangrove. Once absorbed, these cations can cause enzymatic inhibition, generating physiological stress, reduced growth and even desiccation (death) when drought conditions prevail [31]. On the other hand, the tidal level also influences sediment rinsing and with it removal of the excess salts that accumulate in the sediment or in the roots due to evaporation or evapotranspiration of the silt, as has been reported by [32]. Therefore, the high tide level is an important factor inhibiting e salt accumulation in the sediments of mangroves.

The highest temperatures in the reforested zones occurred from June to October which could be attributed to the climatic variability of the region as it coincides with the hottest months and with the onset of the rainy season [25]. During this period, the sediments are exposed to wide temperature (26–35°C) variations between day and night, in addition to decreased water column height, exposing surface sediments to a higher incidence of solar radiation. These variations were recorded at Basin III (3E, 3F, 3G, 3H) stations, with the highest temperatures recorded during June to October. On the other hand, the temperatures were lower in stations of Basins I and II (1A, 1B and 2C, 2D), due to substrate flooding by the tidal processes which promote water exchange through the Tepalcates and Ventanas Channels. In addition, the tide level is an important factor in the Cuyutlán Lagoon; generally, high-water levels occur in the winter months, decreasing in April due to the spring tides [33] influencing the temperature variations in the lagoon system substrate. Ref. [34] studied the dynamics of mangroves in Portete Bay (Cuba) and reported that temperature contributes to the zonation of different (plant) species of mangrove in coastal ecosystems, while Ref. [35] indicated that the temperature inside these systems (lagoons or estuaries) is an important factor at the physiological level, regulating growth, photosynthetic processes, respiration, and salt segregation from the organism. On the other hand, Ref. [36] reported that the incidence of radiation is a determining factor for the growth of mangrove ecosystems.

The mangrove reforestation zones showed marked differences in physicochemical parameters ($p < 0.05$) between basins; in general, low oxygen concentrations were observed in Basin III (3E, 3F, 3G and 3H), while elevated values were recorded in Basins I and II (1A, 1B and 2C, 2D). High concentrations of dissolved oxygen were recorded during August to October (mean 3.4 mg/l) associated with water exchange with the Pacific Ocean through the Tepalcates and Ventanas Channels. In addition, rainfall and fluvial contributions to the lagoon generated sediment turbulence, atmospheric aeration, and a subsequent diffusion towards the interior of the sediments [25]. Therefore, the oxygen concentrations of interstitial waters in Basins I and II are associated with circulation mediated by tides [37], which causes turbulence and resuspension of the sediments. In the internal stations of Basin III, no direct influence of the tide was registered and low values of dissolved oxygen were observed. Ref. [38] carried out a work in the bay of Matanzas, Cuba, which reported how the dynamics and topography associated with the bay influence the high concentrations of oxygen because they determine the penetration of the tide and thus the exchange of surface water.

The pH in the interstitial waters of the reforestation zones ranged from 7.7 to 8.4 in the three basins during the study. Low values of pH in in sediments can be attributed to the high amounts of organic matter, which, when decomposed, generate fatty acids and release ions H^+ that are absorbed by the sediments. On the other hand, basic conditions are generally caused by the increase of salts in the sediments which favours the formation of oxidrile ions (OH^-) [39]. Ref. [40] attributed the degradation of organic material in mangrove sediments to intermediate pH values (7–7.5). This increases nutrient availability, because the redox potential of the sediments can increase to positive values (typical of anoxic systems), favouring the presence of organisms associated with mangrove communities, such as polychaetes and bivalves. The activity of these organisms generates diffusion and water exchange, which helps maintain an adequate pH for these species (7–7.5) and results in positive cyclic oxic conditions for the ecosystem.

All the reforested mangrove sampling sites presented variable OC percentages associated with sediment depth. The elevated percentage of organic carbon at the surface can likely be attributed to the high contributions of leaf litter. After accumulation, the remineralisation of the matter generates the formation of new organic compounds, which in turn promote the early diagenesis and later migration of labile OM towards the deep sediments [41, 42]. The increase in organic carbon in the first 5 cm of sediments in Basins I and II could likely be attributed to high rates of sediment accumulation (with high organic matter content) registered in these zones which is influenced by the connections with Manzanillo Bay (Ventanas Channel) and the Pacific Ocean (Tepalcates Channel), respectively [25]. In addition, water from anthropogenic activities and runoff carry new sedimentary material, which rapidly remineralises with the recently deposited organic matter, promoting carbon accumulation below the most active diagenetic zone. Otherwise, the vertical profile shows a decrease of organic carbon through the sedimentary column, which can be interpreted as a characteristic process, because in sediments with high interstitial water flow, carbon is highly remineralised into carbohydrates and lignin [25]. Ref. [43] mentioned that the decreases of the organic carbon are due to a process of bacterial degradation of the first order, which quickly degrade the organic compounds by means of the available oxygen.

Ref. [44] observed that the percentages of organic carbon are negligible at depths of 30–90 cm due to their degradation in the presence of dissolved oxygen. This pattern was seen in Basin III stations, which had a 30% reduction of organic carbon in the first 5 cm. Ref. [45] suggested that the presence of organic carbon in sediments associated with mangroves controls the variations of the physicochemical parameters (salinity and pH) in the sediment column, due to the leachates that are generated. In this work, percentage organic carbon (2–15%) in the sediment is in the range reported by [46] in the most important mangrove ecosystem of the Gulf of Mexico (Tabasco). This suggests that the diagenetic process found in this system is in agreement with the availability of organic material and the sediment stability of a tropical coastal system.

The predominance of the MS fractions at the sampling points where there was no mangrove reforestation can in all likelihood be attributed to the sediment runoff, due in the first instance to the direct contact between the stations with the sandbar that divides the lagoon from the Pacific Ocean, and in the second, the proximity of the Basin II to the Tepalcates Channel which promotes the accumulation of inorganic sedimentary material by the high hydrodynamics associated with this area [25]. Similarly, the effect of the terrestrial runoff into the lagoon is a determining factor for the accumulation of these sediment fractions. Similar trends have been reported by [5] who found that MS and coarse sands predominate in areas with greater dynamics; while Ref [47] conclude that the topography favours distribution, because coastal lagoons with greater depth favours the presence of fine inorganic sediments and in shallower areas, thicker sizes (coarse sands). Regarding the relationship between grain size and mangrove growth, Ref. [16] mention that in these ecosystems, the main sedimentary composition associated with the best development of the mangrove are fine sediments (silt-clays) that favour the accumulation of more material organic, avoiding its early degradation, contrary to sediments that generate interstitial water penetration and therefore, a high rate of degradation of accumulated organic matter.

http://dx.doi.org/10.5772/intechopen.73045

In the coarse sand sedimentological fraction, the dominant factor contributing to the variation in e porosity was the biogenic composition (bioclasts), which generated high porosities. At depth, the MS fraction dominated favouring a smaller interstitial space between the sediments. On the other hand, the stations of Basin III (3E, 3F, 3G and 3H) show homogeneity in the sedimentological fractions (MS) so they have relatively low porosities, this pattern can be attributed to the benthic vertical flow within the sediments. Ref. [25] demonstrated that the accumulation of new sedimentary material compacted the deeper sediments resulting in elevated porosities only in the upper surface layers. This was observed at the stations of Basins I (1A and 1B) and II (2C and 2D) during February where flooding and proximity to urbanised areas, which contributed to new anthropogenic material with coarse size due to continental runoff, increasing interstitial spaces; and the second is related to the contribution of mangrove organic matter and remains of organisms that lead to the accumulation of organic carbon in the form of coarse particles that increase the porosity values of the sediments.

The humidity content of sediment is directly related to the grain size and high porosities, since if the interstitial spaces are larger in the sedimentary column, the accumulation of water is favoured, which favours the diffusive flow of compounds and the degradation of organic matter caused by the presence of oxygen availability to carry out this process. While low humidity content with high porosity are related to the level of tide and the thickness of the water layer above each site, which generates sediments with low or high-water contact affecting the flow inside the sediments. This process was observed in the seasons of Basins I (1A and 1B) and II (2C and 2D) during the month of March where the tide level was low; however, high porosities and low moisture content were recorded.

5. Conclusions

Two principal factors which influenced the physicochemical parameters that support ideal ranges for mangrove development are the morphology of the study area and tidal levels. The decrease in OC with depth relates to bacterial remineralisation which uses the dissolved oxygen resulting from interstitial water exchange; low OC is a stressor for nutrient uptake and, in turn, reduces growth in these types of plants. Mangroves in areas with more favourable hydrodynamics conditions and physicochemical parameter stability, such as those recorded in Basins I and II, were found to be in a better overall condition after replanting. Basins II and II are therefore feasible areas for the conservation of these ecosystems, as opposed to Basin III, where dwarfism (weak mangrove) was registered. Finally, the ecological importance of mangrove ecosystems as filters that increase nutrient concentrations and other sources of organic matter was confirmed.

Acknowledgements

We want to recognise the Federal Electricity Commission (CFE) of Mexico for having allowed us to use the results of this program for the Bachelor's thesis of Carlos A. Zenteno-Palma, from which this work emerges, as well as the students of the Faculty of Marine Sciences of the University of Colima because without them this work would not have been possible. Finally, a special mention to the Olivos Foundation that covered all the extra expenses for this and other related works.

Author details

Carlos Augusto Zenteno-Palma[1], Aramis Olivos-Ortiz[2*], María del Carmen Álvarez[2], Sonia Isabel Quijano-Scheggia[2] and Gloria Alicia Jiménez-Ramón[2]

*Address all correspondence to: aramis.olivos@gmail.com

1 School of Marine Sciences, University of Colima, Manzanillo, Colima, México

2 University Center for Oceanological Research, University of Colima, Manzanillo, Colima, México

References

[1] Lankford RR. Coastal lagoons of Mexico their origin and classification. In: Estuarine Processes: Circulation, Sediments, and Transfer of Material in the Estuary; 1997. pp. 182-215

[2] Kjerfve B. Coastal lagoons. In: Elsevier Oceanography Series. Amsterdam: Elsevier; 1994;**60**:1-8

[3] Kennish MJ, Paerl HW. Coastal Lagoons: Critical Habitats of Environmental Change. Boca Raton, FL: CRC Marine Science Series, CRC Press; 2010

[4] Lara-Domínguez AL, Contreras-Espinosa F, Castañeda-López O, Barba-Macías E, Pérez-Hernández MA. Lagunas costeras y estuarios. La biodiversidad en Veracruz: Estudio de caso. Comisión para el Conocimiento y Uso de la Biodiversidad (CONABIO), Gobierno del Estado de Veracruz, Universidad Veracruzana (UV), Instituto de Ecología AC (INECOL); 2011. pp. 301-317

[5] Calva-Benítez L, Torres-Alvarado R. Carbono orgánico y características texturales de sedimentos en áreas del pasto marino Thalassia testudinum en ecosistemas costeros del sureste del golfo de México. Universidad y ciencia. 2011;**27**(2):133-144

[6] Díaz-Vargas M, Elizalde Arriaga EE, Quiróz Castelán H, García Rodríguez J, Molina Estudillo I. Caracterización de Algunos Parámetros Físico Químicos del Agua y Sedimento del Lago Zempoala. Morelos, México: Acta Universitaria. 2005;**15**(2):57-65

[7] Albertelli G, Covazzi-Harriague A, Danovaro R, Fabiano M, Fraschetti S, Pusceddu A. Differential responses of bacteria, meiofauna and macrofauna in a shelf area (Ligurian Sea, NW Mediterranean): Role of food availability. Journal of Sea Research. 1999;**42**(1):11-26

[8] Guiñez M, Valdés J, Sifeddine A. Variabilidad espacial y temporal de la materia orgánica sedimentaria, asociada a la Zona de Mínimo Oxígeno (ZMO), en un ambiente costero del norte de la corriente de Humboldt, bahía de Mejillones, Chile. Latin American Journal of Aquatic Research. 2010;**38**(2):242-253

[9] López Jiménez MA, Monks S, Serrano A, Pulido Flores G, Gaytan Oyarzun JC, López Ortega M. Dinámica de las variables fisicoquímicas del sedimento de la laguna de Tampamachoco, Veracruz, México. Revista Científica UDO Agrícola. 2012;**12**(4):965-972

[10] Calva-Benítez L, Pérez-Rojas A, Márquez A. Contenido de carbono orgánico y textura de sedimentos del sistema lagunar Chantuto-Panzacola, Chiapas. Hidrobiológica. 2006;**16**(2): 127-136

[11] Márquez B, Jiménez M. Moluscos asociados a las raíces sumergidas del mangle rojo Rhizophora mangle, en el Golfo de Santa Fe, Estado Sucre, Venezuela. Revista de Biología Tropical. 2002;**50**(3-4):1101-1112

[12] Pizarro F, Piedra L, Bravo J, Asch J, Asch C. Manual de Procedimientos para el Manejo de los Manglares de Costa Rica. Heredia, CR: EFUNA; 2004

[13] Restrepo-Martínez J, Vivas-Aguas LJ. Corporación Autónoma Regional del Cauca, C. I. d. I. M. y. C. Manual metodológico sobre el monitoreo de los manglares del Valle del Cauca y fauna asociada, con énfasis en aves y especies de importancia económica (piangua y cangrejo azul). Serie de Publicaciones Generales (INVEMAR). Colombia: Santa Marta; 2007;(21):40

[14] Del Mónaco C, Giménez E, Narciso S, Alfonso F, Bustillos F. Caracterización de los bosques de manglar y las praderas de Thalassia testudinum de la Isla La Tortuga y cayos adyacentes, Venezuela. Boletín del Centro de Investigaciones Biológicas. 2010;**44**(3):297-316

[15] Manrow Villalobos M, Vílchez B. Estructura, composición florística, biomasa y carbono arriba del suelo en los manglares Laguna de Gandoca y Estero Moín, Limón, Costa Rica. Revista Forestal Mesoamericana Kurú. 2012;**9**(23):1-18

[16] Laffoley D, Grimsditch GD, editors. The Management of Natural Coastal Carbon Sinks. Gland, Switzerland: IUCN; 2011. 53 pp

[17] Werner A, Stelzer R. Physiological responses of the mangrove *Rhizophora mangle* grown in the absence and presence of NaCl. Plant, Cell & Environment. 1990;**13**(3):243-255

[18] Díaz Gaxiola JM. Una revisión sobre los manglares: Características, problemáticas y su marco jurídico. Importancia de los manglares, el daño de los efectos antropogénicos y su marco jurídico: Caso sistema lagunar de Topolobampo. Ra Ximhai. 2011;**7**(3):355-369

[19] Monroy-Torres M, Flores-Verdugo F, Flores-de-Santiago F. Crecimiento de tres especies de mangle subtropical en respuesta a la variabilidad en el hidroperiodo en un tanque experimental. Ciencias marinas. 2014;**40**(4):263-275

[20] López Portillo J, Ezcurra E. Los manglares de México: una revisión. Madera y bosques. 2002;**8**(Es1):27-51

[21] Zaldívar-Jiménez A, Silveira-Herrera J, Coronado-Molina C, Alonzo-Parra D. Estructura y productividad de los manglares en la reserva de la biosfera Ría Celestún, Yucatán, México. Madera y Bosques. 2004;**10**(2):25-35

[22] Téllez-García CP. Caracterización dasonómica del bosque de mangles en el estero Palo Verde, Laguna de Cuyutlán. Colima: Tesis; 2010. 77 p

[23] Yong Y, Baipeng P, Guangcheng C, Yan C. Processes of organic carbon in mangrove ecosystems. Acta Ecologica Sinica. 2011;**31**(3):169-173

[24] Sosa-Avalos R, Gaxiola-Castro G, Olivos-Ortiz A, Silva-Iñiguez L. Nutrientes inorgánicos y producción del fitoplancton en una laguna costera subtropical de México. Revista de biología marina y oceanografía. 2013;**48**(1):143-154

[25] Orozco-Duran A, Olivos-Ortiz A, Meyer-Willerer A, González-Chavarin I. Carbono Orgánico, nitrógeno total y diagénesis en sedimentos superficiales de la Laguna de Cuyutlán. In: Investigaciones Marinas y Acuícolas sobre el Pacífico Tropical Mexicano Colima, México. 2010;**1**(1):248

[26] Jiménez-Ramón GA, Espejel MI , Silva-Iñiguez L, Chávez-Comparan JC, Lara-Chávez B, Pérez-López R, Ortega Ortiz C, Nebereta A, Diaz-Torres E, Torres M, Arredondo y C. Leyva. Indicadores de Monitoreo de Impacto en Playas, Puertos y Humedales de la zona costera de Manzanillo. 1 Congreso Iberoamericano de Gestión Integrada de Áreas Litorales. Mirando a Iberoamérica. Libro de Comunicaciones y Posters. Cádiz (España); Colima, México. 2012. pp. 1209-1221

[27] American Public Health Association, American Water Works Association (APHA-AWWA). Standard Methods for the Examination of Water and Wastewater: Selected Analytical Methods Approved and Cited by the United States Environmental Protection Agency. Washington, DC: American Public Health Association; 1981

[28] Gutiérrez-Estrada M, Galaviz-Solis A, Castro-Del Rio A. Variación de la textura, densidad, porosidad y otras propiedades físicas de los sedimentos del talud continental al SW de la península de Baja California, México. Anales del Instituto de Ciencias del Mar y Limnología-UNAM. 1988;**15**(2):9-21

[29] Cabral SEG. Efectos antropogénicos sobre la calidad del agua, diversidad y abundancia de la fauna nectónica de la laguna de Cuyutlán, Colima, México: recomendaciones para su manejo. Tesis CIBNOR Esp. Doctor en Ciencias en el Uso, Manejo y Preservación de los Recursos Naturales (Ecología) Baja California Mexico; 2011

[30] Odum WE, McIvor CC, Smith TJ III. The Ecology of the Mangroves of South Florida: Community Profile US Fish and Wildlife Service. Washington, DC: Office of Biological Services FWS/OBS-83/17; 1982

[31] Prieto Méndez J, González Ramírez AD, Prieto García F. Contaminación y fitotoxicidad en plantas por metales pesados provenientes de suelos y agua. Tropical and Subtropical Agroecosystems. 2009;**10**(1):29-44

[32] Flores Verdugo F, Moreno Casasola P, Agraz Hernández C, López Rosas H, Benítez Pardo D, Travieso Bello A. La topografía y el hidroperiodo: dos factores que condicionan la restauración de los humedales costeros. Boletín de la Sociedad Botánica de México. 2007;**80**(Suppl):33-47

[33] Mellink E, Riojas-López ME. Modificaciones estructurales artificiales de Laguna Cuyutlán, Colima, México. Revista Geográfica. 2007. pp. 131-142

[34] Molina Grajales EC. Dinámica de los manglares de Bahía Portete, alta Guajira a escala de paisaje y su relación con variables climáticas asociadas al cambio climático global y regional. Tesis de la Universidad Nacional de Colombia, Sede Medellín; 2009

[35] Waycott M, McKenzie LJ, Mellors JE, Ellison JC, Sheaves MT, Collier C, Schwarz AM, Webb A, Johnson JE, Payri CE. Vulnerability of mangroves, seagrasses and intertidal flats in the tropical Pacific to climate change. In: Bell D, Johnson JE, Hobday JA. editors. Vulnerability of Tropical Pacific Fisheries and Aquaculture to Climate Change. Secretariat of the Pacific Community, Noumea, New Caledonia. 2011. pp. 297-368

[36] Febles-Patrón JL, Novelo J, Batllori E. Efecto de factores abióticos en el desarrollo de raíces primarias, crecimiento y supervivencia de propágulos de *Rhizophora mangle* L. Revista Madera y Bosques. 2007;**13**(2):15-27

[37] Gaviño Rodríguez JH, Galicia Pérez MA, Olivos Ortiz A. Hidrodinámica de la laguna de Cuyutlán. In: CD: Resúmenes XIV Congreso Nacional de Oceanografía Manzanillo, Colima. Mayo 2006. pp. 15-19

[38] Almeida Rodríguez M, de Armas Vargas Y, Rojas López L. Caracterización química de los sedimentos sumergidos, elemento clave dentro de un programa de monitoreo ambiental integral en las playas de la bahía de Matanzas. Cuba: Oficina de Manejo Costero-Playa de Varadero; 2010

[39] Huerta HEC. Determinación de propiedades físicas y químicas de suelos con mercurio en la región de San Joaquín, Qro., y su relación con el crecimiento bacteriano. Tesis de licenciatura en Biología. Querétaro, México: Universidad Autónoma de Querétaro; 2010

[40] de Jesús-Navarrete A. Distribución de los sedimentos en la Bahia de Chetumal, Quintana Roo, Mexico. Hidrobiológica. 2000;**10**(1):61-67

[41] Alongi D, Wattayakorn G, Pfitzner J, Tirendi F, Zagorskis I, Brunskill G, Davidson A, Clough B. Organic carbon accumulation and metabolic pathways in sediments of mangrove forests in southern Thailand. Marine Geology. 2001;**179**(1):85-103

[42] Alongi DM. Carbon sequestration in mangrove forests. Carbon Management. 2012;**3**(3):313-322

[43] Berner RA. Early Diagenesis: A Theoretical Approach. Princeton, N.J: Princeton University Press; 1980

[44] López P, Vidal M, Lluch X, Morguı JA. Sediment metabolism in a transitional continental/marine area: The Albufera of Majorca (Balearic Islands, Spain). Marine and Freshwater Research. 1995;**46**:45-53

[45] Marchand C, Baltzer F, Lallier-Vergès E, Albéric P. Pore-water chemistry in mangrove sediments: Relationship with species composition and developmental stages (French Guiana). Marine Geology. 2004;**208**(2):361-381

[46] Moreno Cáliz E, Peña AG, Castorena MdCG, Solorio CAO, López DJP. Los manglares de Tabasco, una reserva natural de carbono. Ed. A.C: Instituto de Ecología; 2002;**8**(Es1):115-128

[47] Escobar-Briones E, García-Villalobos FJ. Distribution of total organic carbon and total nitrogen in deep-sea sediments from the southwestern Gulf of Mexico. Boletín de la Sociedad Geológica Mexicana. 2009;**61**(1):73-86